Spaces of Sustainability

The turn of the millennium will be remembered for its opulent celebrations and underlying anxieties, but perhaps its most enduring legacy will be the prevailing concerns which were expressed at the time about humanity's long-term social, economic and environmental sustainability. Climate change, global poverty, chronic water shortages, economic instabilities and dwindling supplies of non-renewable resources all appeared to coalesce at the end of the twentieth century to cast a spectre over what the twenty-first century would hold. *Spaces of Sustainability* offers a unique insight into contemporary attempts to create a more sustainable society.

Spaces of Sustainability provides an accessible introduction to the key philosophical ideas which lie behind the principles of sustainable development. This book introduces the debates surrounding sustainable development through a series of interesting case studies. These include discussions of land-use conflicts in the USA; agricultural reform in the Indian Punjab; environmental planning in the Barents Sea; community forest development in Kenya; transport policies in Mexico City; and political reform in Russia. Rather than depicting the search for a more sustainable future as a historically unfurling and inevitable process, this book uncovers the tense geographical struggles which surround and inform emerging brands of sustainability. Travelling through Africa and North America, Latin America and Western Europe, Asia and the Arctic, *Spaces of Sustainability* reveals the existence of multiple, politically and culturally specific forms of sustainability. In addition to exposing the varied geographies of sustainability, analysis also considers how sustainable development is being pursued through a range of newly emerging scales of social and economic organization including community spaces, sustainable cities, ecological regions and global agreements.

Written in an engaging and conci~ ~~~~~~~~~~~~~~~~~~~~~~
essential reading for students of geogr
tics and urban studies. It is illustrated
along with a range of explanatory hel

Mark Whitehead is a lecturer in hu~~~~~ ~~~graphy at the Institute of Geography and Earth Sciences at the University of Wales, Aberystwyth.

Frontispiece A sustainable telephone box at the Centre for Alternative Technology (Wales)

Spaces of Sustainability

Geographical perspectives on the sustainable society

Mark Whitehead

Routledge
Taylor & Francis Group

LONDON AND NEW YORK

First published 2007
by Routledge
2 Park Square, Milton Park, Abingdon, Oxon OX14 4RN

Simultaneously published in the USA and Canada
by Taylor and Francis Inc.
270 Madison Ave, New York, NY 10016

*Routledge is an imprint of the Taylor & Francis Group,
an informa business*

© 2007 Mark Whitehead

Typeset in Times New Roman and Franklin Gothic by
Florence Production Ltd, Stoodleigh, Devon
Printed and bound in Great Britain by
Antony Rowe Ltd, Chippenham, Wiltshire

British Library Cataloguing in Publication Data
A catalogue record for this book is available from the British Library

Library of Congress Cataloging in Publication Data
Whitehead, Mark, 1975–
    Spaces of sustainability: geographical perspectives on the
    sustainable society/Mark Whitehead.
        p. cm.
    Includes bibliographical references and index.
    1. Sustainable development – Case studies.   I. Title.
HC79.E5W485 2007
338.9′27–dc22   2006015005

ISBN10: 0–415–35803–5   ISBN13: 978–0–415–35803–3 (hbk)
ISBN10: 0–415–35804–3   ISBN13: 978–0–415–35804–0 (pbk)
ISBN10: 0–203–00409–4   ISBN13: 978–0–203–00409–8 (ebk)

In memory of
Gwen, Jim and Vaughan Whitehead.

You are loved and missed
in equal measure.

CONTENTS

FIGURES AND TABLES

Figures

Tables

BOXES

ACKNOWLEDGEMENTS

Hidden between the lines and pages of this book are innumerable traces of the support and help I have received during its production. First, I would like to acknowledge the staff of the Institute of Geography and Earth Sciences, at Aberystwyth, for helping to create such a supportive and enjoyable place to think and write about geography. I would like to thank especially Bill Edwards and Martin Jones for their continued guidance and wise words, and Luke Desforges, Deborah Dixon, Kate Edwards, Graham Gardner, Gareth Hoskins, Peter Merriman, Heidi Scott and Mike Woods for their valued friendship. I would also like to express my gratitude to Bob Dodgshon, who first gave me the idea of writing a book exploring the links between geography and sustainability; Rhys Jones, who, in addition to coming up with the title for this book, has been a valued source of humour and distraction; and Tim Cresswell, whose own work and style of writing have inspired my endeavours in this volume. Beyond my own department I must also acknowledge the guidance and input of Marcus Doel, Craig Johnstone, Mark Goodwin and Gordon MacLeod. A mention must also be made of my exceptional group of graduate students – Anna Bullen, Keiko Nakamura, Julie MacLeavy, Kelvin Mason, Ruth Stevenson and Chris Yeomans – who through their research and writings have introduced me to a range of new issues relating to sustainability and challenged the way that I understand the notion of a sustainable society. Thanks also go to the numerous undergraduates here at Aberystwyth with whom I have shared the ideas expressed in this volume. As ever I am grateful to my family (and in particular Janet, Patrick and Louisa Davies and Andrew and Carol Hawkins) for their continued interest in my work and their enthusiastic enquiries into how 'the book' was going – I love you all. Finally, I would like to say thank you to Sarah, your love has, as ever, enabled me to do what I do, as best as I can possibly do it. This book is dedicated to the memory of my Father and Grandparents.

All photographs are from my own collection, except for Figure 5.3 which is reproduced courtesy of Michael Hambrey.

ACRONYMS AND ABBREVIATIONS

BRT   bus rapid transport
CAT   Centre for Alternative Technology
CIAT   Centro International Agriculture Tropical
CST   Center for Sustainable Transport in Mexico City
DEFRA   Department for Environment, Food and Rural Affairs (UK)
DETR   Department of the Environment, Transport and the Regions (UK)
DoE   Department of the Environment (UK)
EJM   Environmental Justice Movement (USA)
EPA   Environmental Protection Agency (USA)
GBM   Green Belt Movement (Kenya)
GDP   gross domestic product
HYVs   high-yielding varieties
IBP   International Biological Programme
ICLEI   International Council for Local Environmental Initiatives
ICSU   International Council of Scientific Unions
IITA   International Institute of Tropical Agriculture
IMF   International Monetary Fund
IPCC   Intergovernmental Panel on Climate Change
IUBS   International Union of Biological Sciences
IUCN   International Union for the Conservation of Nature and Natural Resources (World Conservation Union)
LA21   Local Agenda 21
LEDCs   less economically developed countries
MEDCs   more economically developed countries
NAFTA   North American Free Trade Agreement
NGO   non-governmental organization
PCBs   Polychlorinated biphenyls
PLF   Pacific Legal Foundation
SKAP   Sustainable Katowice Agglomeration Project
UNCED   United Nations Conference on Environment and Development

UNCHE   United Nations Conference on the Human Environment
UNCHS   United Nations Centre for Human Settlements
UNEP   United Nations Environment Programme
WCED   World Commission on Environment and Development
WHO   World Health Organization
WWF   World Wildlife Fund

1 THE GEOGRAPHIES OF THE SUSTAINABLE SOCIETY

While I was writing sections of this book I spent a brief time living and working in London. During this spell my morning walk in to work took me along the bustling Cromwell Road and past the world famous Natural History Museum. Although I had little time to stop, every morning I would catch glimpses of an outdoor exhibition at the museum, which was comprised of enormous colour photographs of different parts of the world, which had been taken from the air. Even though I used to enjoy looking at those exotic photographs in the morning I never did make (or indeed have) the time to stop and look at the exhibition more closely. Later that year I was walking through my home city of Birmingham when I saw a now familiar set of photographs laid out on the streets and squares of the city. The exhibition I had first seen in London it transpired was on tour. It was now clear to me that I was destined to look at it properly – I wasn't disappointed. The exhibition was actually entitled the Earth from the Air and was a collection of 160 photographic images by the famous French photographer Yann Arthus-Bertrand. The exhibition represented the culmination of a UNESCO-sponsored project to produce a 'photographic record of the natural world at the start of the new millennium'. What had been produced was a beautiful array of photographs, covering all of the Earth's continents, and combining views of natural landscapes, cityscapes and agricultural areas. I found the exhibition compelling (I actually viewed it over three days) for two reasons: first because of its sublime beauty and intrinsic fascination; and second because of the way it seemed to speak directly about the links between sustainability and geography I am interested in.

To understand the reasons why this exhibition seemed to speak so directly about the links between sustainability and geography, it is important to reflect upon the intentions of the artist/photographer Yann Arthus-Bertrand. While wishing to produce a photographic record of the world at the turn of the millennium, Yann Arthus-Bertrand was also keen to show how human development was threatening the social and environmental fabric of the planet. In this context, Arthus-Bertrand juxtaposed images of

Figure 1.1 People exploring the Earth from the Air Exhibition – Victoria Square, Birmingham (July, 2004)

the fragile natural beauty of the Earth (including images of Mount Everest; a caravan of dromedaries in Mauritania; atolls in the Maldives; and glaciers in southern Argentina) with examples of the huge socio-ecological impact that humans are having on the planet (including panoramas of Tokyo and New York cities; an Iraqi tank graveyard in Kuwait; and a deserted residential district surrounding the Chernobyl nuclear power plant). In addition to visual imagery though, throughout the exhibition viewers were presented with statistics relating to rates of environmental destruction, pollution growth and adult literacy in Africa. On other boards you could find information about the principles of sustainable development – an idea which Arthus-Bertrand clearly sees as being of critical importance in addressing the socio-ecological issues he is attempting to raise. A further intention of Yann Arthus-Bertrand, however, was that others could continue the work of this exhibition, by taking their own photographs. Arthus-Bertrand assisted viewers in this exercise by providing details of the photographic techniques and methods he used and by giving precise coordinates for each of the 160 images he had taken, so that others could return there. My favourite photograph for example was of a crowd of people in Abengourou, Côte d'Ivoire, which was taken at precisely N 6° 44', W 3° 29'.

To me then, this enthralling exhibition was nothing less than a visual representation of the themes I intend to cover in this book – it was a manifestation of the geographies of the sustainable society. I say this not simply because the exhibition was obviously geographical (it was a collection of photographs from around the world) and because one of its stated aims was to raise awareness about the need for sustainable development. To me the exhibition provided a geographical perspective on the sustainable society because, when taken as a whole, the 160 images that the exhibition brought together served to illustrate the global nature of the issues which the creation of a more sustainable society must address. Added to that, when approached individually, each photograph of a unique place (perhaps N 6° 44′, W 3° 29′) served to illustrate the very particular ways in which the social, economic and environmental processes that determine whether something is sustainable are constituted within different geographical circumstances.

This volume interprets sustainability in two basic ways: first in terms of the long-term durability of human, ecological and economic systems; and second on the basis of how different human, ecological and economic systems interact in order to determine their own relative levels of sustainability. Geographers have historically had a very keen interest in the notion and implications of sustainable development. Since the popularization of the term in the late 1980s, the idea of sustainability has interested both human and physical geographers alike. Indeed, from the late 1980s onwards, a number of pioneering writings started to appear in geographical books and journals on issues of sustainability. Some human geographers for example sought to build on work exploring the links between environmental perception and behavioural geography to understand how sustainable development, as a discourse, may be transforming human understandings of the environment (see Burgess 1990: 139)[1]. Perhaps the leading field in early geographical work on sustainability was development geographies. During the late 1980s to early 1990s there was a proliferation of geographical writings on the tensions over sustainable development in less economically developed countries (see Adams and Hollis 1989; Stocking and Perkin 1992). These types of study culminated in W.M. Adams now classic text, *Green Development* (1990b), which explored sustainable development issues in a range of different development contexts (see also now Adams 2001). At around this time, it is also possible to discern an emerging neo-Marxist inflected critique of sustainability within key geographical texts (see for example O'Riordan 1989). Throughout the 1990s, geographers' concern with sustainability grew, as physical geographers began to consider the likely climatic effects of sustainable development policies (Warrick 1993), while others considered the spatial planning implications of sustainability (see Thomas and Adams 1997; Cowell 1997; Owens 1994). More recently geographers have been exploring the application of sustainability within a variety of different urban, regional and rural contexts in the developed world (Haughton and Counsell 2004; Marsden *et al.* 2001; Munton 1997a; Whitehead 2003a), the rise of sustainable development within post-socialist transitions (Oldfield 1999, 2001; Pavlínek and Pickles 2000), and the spread of sustainable development thinking within the business world (Eden 1994). Some have even started to suggest that a concern with sustainability could provide a basis for bridging the divide between human and physical geography (Richards 2004). Beyond those who can be officially 'labelled' as geographers, however, it is important to recognize that a number of writers from beyond the geographical discipline have also begun to explore the geographical dimensions of sustainability (see for example Becker and Jahn 1999; Pearce *et al.* 1990; Yearley 1996).

Despite the emerging historical relationship between geographical study and sustainability over the last twenty years (for reviews see Eden 2000; Munton 1997b), it is still unclear what a geographical perspective on sustainability involves, or how this perspective may or may not be different to other disciplinary approaches to sustainable development. This chapter has two primary goals. First, it serves to introduce the idea of sustainability and the related vision of a sustainable society. Second, the chapter offers one explanation of what a geographical perspective on the sustainable society involves, and why such a perspective is important. The first section thus considers what a spatial perspective on the world offers sustainability studies and how contemporary patterns of globalization are transforming understandings of sustainable space. The following section then explores how the creation of a more sustainable society has been driven by fears over different socio-ecological dystopia. The penultimate section of this chapter then analyses the philosophical and geopolitical origins of sustainability and considers its implications for how we organize space. By way of conclusion, this chapter closes by reasserting what geography can offer to sustainability studies.

Geography, globalization and the sustainable society

In order to understand the value of a geographical perspective on the sustainable society, it is important to be aware of the recent contributions which geography, as a disciple, has made to the broader social sciences (see Soja 1989: chapter 1). Soja argues that geography has played a central role in the recent *reassertion* of space into analyses of the social, economic, political and cultural worlds within which we live (ibid.). In this sense a geographical perspective on the sustainable society is, at least partially, about developing a spatial perception of sustainability. This spatial perception is in part about understanding how the construction of more sustainable societies is related to the production of new spatial systems within which people can live, work, commute and interact with the natural world in more sustainable ways. At perhaps a more profound level, however, this spatial perspective on sustainability challenges how we understand and interpret sustainability as a concept. For me then, a geographical perspective on the sustainable society questions the conventional accounts of sustainability we often read, which describe it as the outcome of a long historical series of policy developments and negotiations. Such historical accounts often talk about the key meetings, publications and protocols that have all contributed to the gradual evolution of sustainability as a key international policy goal. My problem with these depictions of sustainability is that they tend to (often inadvertently) reduce sustainability to the historical emergence of a singular concept of social and ecological development – that of *sustainable development*. To consider the spatialities of the sustainable society, then, is to become aware of the stories, struggles and values which *cut across* the history of sustainable development (see Soja 1989: 21–24). A spatial perspective on the sustainable society consequently alerts us to the varied cultural and political versions of sustainability which exist alongside, and sometimes in opposition to, prevailing doctrines of sustainable development. This book is dedicated to exploring these geographies of sustainability. Before we begin this task, however, it is important to consider the changing geographical parameters associated with globalization.

Many people's intuitive understanding of geography is that of a discipline which is devoted to studying different societies. On these terms, of course, the idea of society, during the twentieth and twenty-first centuries at least, has been routinely equated with those bounded territories, or states, which are drawn on the political map of the world with which we are all so familiar. This classical geopolitical view of the world is a world of nation states, each with their sovereign powers, national interests and political territories. So-called *realist* accounts of the world in international politics have described this geopolitical system as one which is based upon the selfish interests of individual states, each looking after their own needs and guarding their social and economic assets through military force. It is the assertion of this book that if we are to understand the relationship between geography and the sustainable 'society' we need to challenge these rather limited assumptions about what both geography and societies are.

A number of things have occurred over the last thirty-five years which have challenged the neat political division of the world into self-contained national states. While the challenges to a nation-state oriented view of the world have at times been economic, social and environmental, they have all in one way or another been embroiled within the complex set of processes which we collectively refer to as *globalization*. David Held *et al.* provide us with a useful insight into what these processes of globalization actually are:

> Globalization may be thought of initially as the widening, deep-
> ening and speeding up of worldwide interconnections in all
> aspects of contemporary social life, from the cultural to the crim-
> inal, the financial to the spiritual. That computer programmers
> in India now deliver services in real time to their employers in
> Europe and USA, while the cultivation of poppies in Burma can
> be linked to drug abuse in Berlin or Belfast, illustrates the ways
> in which contemporary globalization connects communities in one
> region of the world to developments in another continent.
>
> (Held *et al.* 1999: 2)

As this quote from Held suggests, new patterns of migration, global
economic transactions and the flow of various forms of electronic information around the world are making the boundaries surrounding nation states
look increasingly meaningless. One issue which is conspicuously absent
from this quote, however, is the role of environmental issues in the
globalization debate (although it is covered in a dedicated section in
the book from which this quote is taken, see Held *et al.* 1999: chapter 8;
see also Yearley 1996). The environment has been a factor within the
processes associated with globalization for three reasons: (1) because of
the increasing incidence of transboundary pollution events; (2) because
of the growing social awareness of systemic forms of environmental
change; and (3) because of a growing scientific consciousness of the integrated nature of environmental systems at a global level (these processes
are all explained and explored in greater detail in Chapter 6). Essentially,
over the last thirty years scientific studies of the environment have shown
that national boundaries make little sense ecologically – ecosystems do
not correspond to political boundaries; acid rain does not only fall within
the borders of the state within which it was produced; global warming is
occurring irrespective of which state you happen to live in.

Within human geography such processes have led to a significant shift
in how social, economic and cultural spaces are understood. At the centre
of this shift has been a desire to understand space not as a bounded set of
fixed locations, but more as bundles of interconnected relationships
(Massey 1994, 2004 on the notion of *relational space*). Consequently,
geography and geographers are becoming increasingly interested in the
spatial connections of global society and the global environment. The
importance of this borderless vision of the Earth to contemporary geography is twofold. First, it clearly emphasizes the emerging geographical
concern with global processes. Second, and perhaps more subtly, however,
it also emphasizes an intensified geographical interest in social, economic
and ecological processes at a sub-national level. If globalization has challenged the integrity of nation states from above (so to speak), it has also

provided the context for the emergence of a whole range of new sub-national spaces – including economic growth regions, global cities and transnational communities, which operate both above and below the scale of the nation. Many writers now claim that it is these spaces of economic and cultural life which are shaping the future of the planet and creating a distinctively post-national world (see Bulkeley 2005). It is in the context of this maelstrom of new, fluid, post/trans/sub/intra/supra-national 'societies' that our discussions of the sustainable society must consequently begin. These discussions of globalization should not, however, lead us to a naive abandonment of nation states within analyses of the sustainable society. Chapters 2, 3 and 4 in this book for example all illustrate that states continue to provide important legal, moral, political and cultural contexts within which different forms of sustainability are emerging. The point is that while states continue to influence the sustainable society, we cannot understand the geographical dimensions of the sustainable society from a state-based perspective alone.

To talk about the sustainable society is consequently to talk about a plurality of different societies and different spatial scales. At one level it is about exploring global society. The idea of a global society has been a popular mantra of the environmental movement since the 1960s as it has consistently emphasized the need to 'think globally', 'to save the planet' and to travel together harmoniously on 'spaceship earth'. Of course the idea of the global, or planetary, society is important not only because of its unifying environmental symbolism, but because it represents a society which cuts across ethnic, political, racial and class divisions and could potentially unite all of humanity through their common inhabitation of 'one world' (although for a taste of the dangers of such visions see Cosgrove 2001). In this context the sustainable society is nothing if it is not a global society. While the global society is obviously important to any under-standing of sustainability, when you stop and think about it, it is actually very difficult to do things globally (at least in the fullest sense of the world). Yes global corporations like Nike, Nestlé or General Motors do have global operations networks, but very little of what they do (apart from perhaps their pollution outputs) can be thought of as having completely global effects. Naomi Klein reflects ironically on the exaggerated myth of global-ization in the opening section of her influential book *No Logo*:

> [t]he euphoric marketing rhetoric of the global village, an incred-ible place where tribespeople in remotest rain forests tap away on laptop computers, Sicilian grandmothers conduct E-business and 'global teens' share [. . .] 'a world-wide culture'. Everyone from Coke to McDonald's to Motorola has tailored their marketing strategy around this post-national vision.
>
> (Klein 2001: xvii)

To Klein, then, the idea of an evenly accessible global experience, or lifestyle, is little more than a marketing strategy used by major corporations to meet their own needs.

The actions of global corporations and political institutions are only global to the extent that they have the potential to influence planetary existence. But it is important to remember the actual impact of such global actions is always locally specific and unevenly developed (the degree of embeddedness within the networks of globalization obviously varies greatly when you compare rainforest tribespeoples with Wall Street bankers for instance). Consequently, and in relation to discussions of the sustainable society, while we may talk of the need for planetary agreements and action, such global actions have to be understood in the context of when, how and where they 'touch down' in geographical space. The critical matter is that issues pertaining to sustainable development touch down in a bewildering multitude of different places, spaces and scales. Some spaces include large ecological areas like the Sahara region in North Africa, which crosses numerous African states and whose social and ecological sustainability is constantly being threatened by the ebb and flow of desertification events. Other spaces include large metropolitan areas like Mexico City, which consume vast quantities of water and natural resources and whose industries disperse large amounts of pollution into the atmosphere. Smaller-scaled societies also matter when it comes to sustainability, like the communities, neighbourhoods or cultural spaces we all inhabit. The cultures which derive from our lived spaces are vital in defining our own everyday contributions to sustainability, whether it is in terms of our patterns of consumption, our recycling habits or our attempts to conserve energy.

It is in this context that I want to argue that the sustainable society is best thought of not as a single, geographically fixed society which can be observed, mapped and analysed (like for example the United Kingdom) but instead as a series of emerging, geographically diverse and inter-connected social spaces and scales. On these terms the study of the sustainable society should be understood as an analysis of the processes which connect different societies and the economies and environments which sustain them. This idea of interconnection is vital here because it suggests the importance of understanding societies, economies and environments less as things (the USA, NAFTA, the Amazonian rainforest) and more as entangled processes and networks (including political power, economic relations, carbon flows). To study the sustainable society geographically is consequently not just about the traditional regurgitation of geographical knowledge concerning different societies (like the USA has a per capita GDP of $37,800; or that the Ethiopian government faces claims of up to $500,000 dollars in debt from forty different claimants),[2] but is instead about understanding the connections (both social, economic and environmental) which exist between different societies (and for that matter these two statistics!). This volume is consequently dedicated to uncovering the diverse spatial strategies through which the principles of sustainability are currently being pursued. It also considers how the very notion of sustainability is leading to the creation of new, interconnected scales of social and political organization (including local communities, cities, regions, states and supra-national networks) which are being used in an attempt to manage the complex social, economic and environmental processes which determine how sustainable our world is (see Bulkeley 2005).

The unsustainable roots of the sustainable society: the age of socio-ecological dystopia

The more observant readers will have already noticed that while I have outlined what I understand the societal aspects of the sustainable society to be (namely multiple, interconnected and poly-scalar social spaces), I have as yet said very little about what I understand about the term sustainability itself. Now for reasons which become clearer later in this chapter, it is not my intention to offer a definitive definition of sustainability. Instead what I propose to do is to describe the context within which discussions of sustainability became possible and some would argue necessary, and the different ways in which the notion of sustainability has subsequently been interpreted and used. Starting from first principles, to be sustainable simply means to enable things to continue to achieve a form of existence which can be maintained indefinitely (the word sustainable actually derives from the Latin word *sustinēre*, which denotes a sense of support in both physical and emotional terms). The word sustainable is actually an adjective, which means it is used to qualify, clarify or add meanings to nouns (or names) and other phrases. Consequently, when we use the term sustainable we rarely use it in isolation, but instead add it to words like agriculture, economy, environment and of course society. In this context the word sustainable transforms the way we understand the world around us and suggests that instead of promoting unmaintainable practices like clear-cutting forestry, boom and bust economics, or environmental pollution, we develop sustainable systems of forestry, a sustainable economy, a sustainable environment. In the contemporary world you may have noticed that the phrase 'sustainable' has become a very fashionable adjective, appearing in an increasingly wide range of contexts (and qualifying an increasingly large number of nouns). Having established this rather simple and rudimentary understanding of the word sustainable, however, perhaps paradoxically, I want to begin a more detailed discussion of sustainability by making reference to things which are unsustainable. While this may seem an unusual thing to do, I feel that a much deeper understanding of what it takes to be sustainable can be gained when you first think of what it is to be unsustainable.

Tales of socio-ecological dystopia

I often think of the sustainable society as a post-industrial utopia. By this I mean that the idea of the sustainable society in part reflects a desire to radically restructure the unsustainable practices which have become synonymous with the modern industrial era. If the sustainable society is a post-industrial utopia, however, it has clearly been driven by some rather bleak dystopic visions of the industrial future. In 1972, for example, the Club of Rome produced its now infamous *Limits to Growth* report (Meadows *et al.* 1972) in which they described the potentially cataclysmic implications resulting from the continued use of non-renewable resources like coal and oil. In classic neo-Malthusian terms, the Club of Rome saw a future which was based upon a series of drastic checks on population growth through either resource shortages or pollution events. Forty years before the *Limits to Growth* report, Aldous Huxley painted an equally bleak vision of the future in his book *Brave New World* (1932) (Huxley 1994). Huxley's book is set in the year 632 AF (After Ford – Henry Ford) and describes a world of mass production and mass consumption initiated through Henry Ford's innovative methods of motorcar manufacture in Detroit. The Brave New World Huxley depicts is, however, a bleak world in which humans are reduced to merely economic agents of production and consumption, who once economically inactive (due to long-term illness or old age) are seen as worthless to society and are subject to widespread programmes of euthanasia. It is a world where social value, meaning and love have been lost – a world which is socially unsustainable without systematic state-sponsored programmes of drug taking, through which the population is made more subservient and pliable. The *Limits to Growth* report and Huxley's *Brave New World* are, in admittedly different ways, dystopic visions of unsustainable futures – futures which are based upon continued industrial expansion and capitalist socio-ecological exploitation.

There are, of course, many contemporary examples of dystopic tales which serve to reiterate the messages of the Club of Rome and Aldous Huxley. In Roland Emmerich's film *The Day After Tomorrow* (2004), cinema goers are confronted with a particularly pessimistic depiction of an unsustainable society which explores the potential environmental consequences of unabated industrial development. In the film, the 'day after tomorrow' is the day after the full effects of global warming have been experienced. The film draws on the scientific theory that global warming (caused by our current dependency on fossil fuels as our major global energy source, see Chapter 6) could cause increasingly traumatic climatic events in order to depict the socio-ecological consequences created by the onset of a new, human-induced ice age. While the scientific theories upon which the film is based are still heavily debated, *The Day After Tomorrow* focuses on the consequences of large-scale polar icecap melting (generated by global warming) on the global climatic system (many scientists argue that the melting of polar ice-sheets could have a significant impact on ocean currents, as the relative balance of salt and fresh water, which is regulated by these sheets, and partly drives these currents, changes). In the film *The Day After Tomorrow*, the director weaves together the interlocking stories of scientists, political leaders and citizens as a rapid climate change event, or shift, strikes the Earth.

While obviously exaggerated for cinematic effect, the consequences of global warming, which *The Day After Tomorrow* depicts, are highly disturbing. The audience are confronted with mega-tornadoes ravaging California; rising sea levels engulfing New York; hail stones the size of footballs striking south-east Asia; and super-cooled tropospheric air instantly freezing people to death in northern Europe. The global climate that emerges as a result of global warming in this film is one which is fundamentally unsustainable for human inhabitation. Cities are razed to the ground and thousands of environmental refugees, forced to leave their homes, flock towards tropical latitudes in order to stay alive. Throughout the film there is a constant message that although mirroring the last major ice age to the hit the Earth (which ended approximately 10,000 years ago, ushering in the period we now refer to as the Holocene), the intense climatic changes depicted in the film are this time not a natural event, but a product of human (or anthropogenic) factors. *The Day After Tomorrow* opened to a very mixed set of responses. There were those who lambasted its sensationalism, while others questioned its scientific accuracy. But there were many in both the environmental movement and scientific community who, while recognizing the film's limitations, welcomed the basic message it contained – that is that current patterns of social and economic development throughout the world carry with them the threat of destabilizing the delicate ecological balances which sustain life on Earth. In the context of this book, I think that *The Day After Tomorrow* is significant because it represents the fears (perhaps justified, perhaps not) which many now have of a fundamentally unsustainable environmental future. While the film may ultimately only be a sensationalized apocalyptic movie, it does provide an insight into the kind of dystopic environmental visions which are influencing contemporary discussions about the need for a more sustainable global society (for more on contemporary geographical interpretations of disaster scenarios see Pelling 2001, 2003).

There is one particularly interesting scene in the film *The Day After Tomorrow* which involves a glaciologist presenting his predictions on climate change to an international conference in snowbound New Delhi! The vice-president of the United States of America interrupts the glaciologist's presentation to point out that the oil-dependent economy of the USA cannot sacrifice current economic growth in order to ward off the hypothetical environmental threat of global warming. The fictitious comments of the US vice-president are, however, symptomatic of the problems which reducing the production of greenhouse gases poses the global political community (see Chapter 6). The problem with reducing the production of greenhouse gases is that it would require a shift away from our current global dependence on the use and burning of oil. This is difficult because industrial economies are quite literally hooked on oil.

Just stop and think for a moment about the importance of oil to the contemporary global economy. At one obvious level oil is used to power industrial economies, providing the fuels which are used in cars and for the production of domestic and commercial energy. But imagine a world without plastic and the bottles, bags, computers and household appliances which are all now routinely made from plastic. Also imagine a world without the synthetic materials most of our clothes now appear to be made from, a world without cosmetics, without carpets, antiseptics, AstroTurf, shampoo, shaving foam, insecticides and so on (Moore 2003). These and many other things are all produced from oil and the petrochemical industries which transform oil into the everyday products which surround people living in the industrialized world. The point is that oil is not only driving and fuelling the world economy, it also constitutes many of the most valuable commodities upon which our contemporary world is based. This is why industrial nations are finding it so hard to give up (or even phase out) oil use and the economic benefits it brings. Beyond the problems of global warming, however, there is of course another fundamental problem with our contemporary economic dependence on oil: put simply, oil is a finite resource and it will run out! Currently, we (as a global society) are consuming an estimated seventy-five million barrels of oil a day. The only known oil and gas reserves that experts are confident will last beyond the next fifty years (many of our lifetimes) are in the Middle East (Harvey 2003a). Our global economic dependence on oil is clearly unsustainable.

Michael Moore (2003) explores the possible consequences of our unsustainable dependency on oil for future generations in his controversial book *Dude, Where's My Country?* In the chapter 'Oil's Well That Ends Well' (chapter 3) Moore describes a dream set in the future (the year 2054) in which he is helping his great-granddaughter complete a school history project exploring the contemporary era's obsession with oil. In Moore's dream, world population has been drastically reduced by the lack of available energy for heating and subsistence. At the same time, those people who are still alive strive to obtain rubbish tip permits which enable them to gather the little plastic that is still available. In the context of this post-oil society, Moore describes to his great-granddaughter his regrets regarding the past – how he would use a gallon of petrol to go and fetch a pint of milk; how he would wrap his plastic waste in plastic bags so that it could be deposited in landfill sites; and how he would use oil burners simply to heat the air outdoors so he and his family could have a late night barbeque (Moore 2003: chapter 3). As with the film *The Day After Tomorrow*, I'm not necessarily suggesting that you believe that Michael Moore's premonition will come true. What it does, however, is serve to emphasize that our current dependence on oil and plastics, when combined with our failure to invest adequately in the search for alternative energy sources and recycling technologies, means that our current economy is unlikely to have a long future. Not only are industrial economies using non-renewable fuel sources, they are also wasting the little energy and raw materials they possess. We appear to be living in what Alvin Toffler (1970) once famously described as a 'throw-away society'. We are currently quite literally throwing away valuable materials which we may never have the opportunity to use again. Toffler actually claimed that this form of throw-away society was going to encounter a *Future Shock*, which would shake it out of its complacency. In the context of current dire predictions about the onset and likely consequences of climate change, however, it now appears that our generation's future shock may actually be related to the ecological consequences of burning oil rather than to its escalating scarcity.

Making the unsustainable sustainable

The reason I chose these two examples of dystopia was because they draw attention to the different ways something can be considered to be unsustainable. So in the case of *The Day After Tomorrow*, we see the production of an unsustainable environmental system, which necessitates a major climatic shift to be rectified, and which in turn has devastating social consequences. In Michael Moore's dream we see the development of an unsustainable economic system which is undermined by its overdependence on oil. Collectively these examples serve to emphasize that the environment, economy and society are key factors in determining whether something is sustainable or unsustainable. But there is more to debates about sustainability than simply asking whether systems are 'environmentally', 'economically' and 'socially' sustainable – sustainability forces us to consider the interactions between environmental, economic and social systems. In *The Day After Tomorrow*, global warming (an environmental problem) is driven by escalating levels of greenhouse gases in the

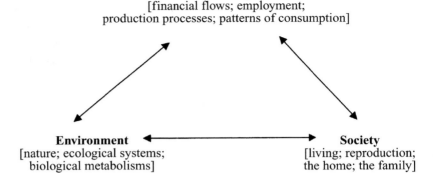

Figure 1.2 Sustainability – an integrated concept

atmosphere (an economic issue), which in turn makes life impossible for those living in areas affected by the changing climate (a social issue). In the case of the oil-based society, while the lack of oil has serious economic consequences, it is socio-cultural practices (driving our cars and using plastic), which continue to make any move away from oil-based products difficult. The global socio-cultural dependence on oil in turn continues to drive significant environmental shifts, which are produced by the pollution which burning oil creates. The point is that issues of sustainability not only make us think about environmental, economic and social issues, but also how these different systems interact, interlock and affect each other. On these terms sustainability is perhaps best thought of as an integrated concept which connects financial flows with human health; consumption with environmental pollution; housing with employment; and ecological systems with modes of production (see Figure 1.2). In this sense it is argued that the key to creating a sustainable society is to develop policies, institutions and ways of life which better reflect the mutual dependency of our environment, economy and society. Throughout this book we see how changing the geographies of how we live is a crucial part of creating a more sustainable environment, economy and society. But where did this idea of sustainability as a solution to global socio-environmental problems come from? The remainder of this chapter sets out a brief history of the philosophical and political ideas which have influenced sustainability.

The birth of a post-industrial utopia: a brief history of sustainability

While it is important at this point to provide you, as the reader, with some sense of where the notion and principles of sustainability have come from, it is not the intention of this book to provide a detailed evolutionary history of sustainable development. The reasons for this are threefold: (1) there already exist a series of books which provide very sophisticated and detailed descriptions of the emergence of sustainability (see Adams 2001; Dresner 2002; Elliot 1999; Kirkby *et al.* 1995; Morris 2002; Redclift 1987); (2) the idea of a history of sustainability is precisely what this book is trying to avoid as it explores the spatial (geographical) rather than the temporal (historical) dynamics of sustainability; and (3) to provide an evolutionary history of sustainability suggests that there is one version of sustainability which has gradually but inevitably emerged over time; as we will see, multiple versions of sustainability coexist in the world and a spatial perspective helps us to recognize these variations. This said there are still certain critical events and developments concerning the emergence of sustainability of which readers of this book need to be aware.

Scientific origins of sustainability: sustainable yields and nature conservation

Redclift (1987: 17–22) observes that many of the antecedents of sustainability as a concept can be discerned within the scientific studies of wildlife and forest management. From the early twentieth century onwards biologists became increasingly aware of the processes of ecological *succession* which acted to secure plant populations' reproduction. Succession is basically the historical process through which various biological species gradually colonize and transform an environment into a stable, interconnected and balanced ecosystem. Successional phases often take a long time to occur as nature gradually builds defences (normally expressed through species diversity) to sudden environmental changes (see Bartelmus 1986: 44). Of course rapid processes of agricultural exploitation can render successional processes redundant and threaten the long-term stability of plant and animal communities. In this context, scientists noticed that if forests and agricultural lands were used in a way which allowed for natural regeneration and succession to occur they could be continually harvested in a sustainable way. The key to the sustainable management of woodland and agricultural resources was the notion of a *sustainable yield*. A sustainable yield is basically an amount, or area of resource (like trees or crop land), that if removed on a particular cycle of time should allow the host ecosystem to recover. In the case of forest management schemes, for example, the principle of sustainable yield would involve only cutting down a certain number of trees, perhaps on a ten-yearly basis, rather than felling the whole forest and destroying the delicate ecosystem upon which it is based (see Dasmann 1985). Even within these early conceptualizations of sustainability there are important geographical issues which need to be acknowledged. First, it is clear that within nascent scientific formulae for sustainability space was a major consideration and basis for calculation. The spatial relations which constitute a successional community, for example, were major factors determining the designation of a process as sustainable or not, while sustainable yields were in part measured on the basis of the spatial extent of a clearance process. Second, it is also crucial to note that the early sciences of resource and forest management were predominantly Western pursuits. It is in this context that many locate the geographical origins of sustainability clearly within more economically developed countries. Suspicions that notions of sustainability are inflected with Western scientific values and ways of approaching the world continue to see sustainability resisted in many parts of the world to this day.

Significantly, in the context of the links between sustainability and ecological conservation, Adams (1990b: 42–51) claims that the first clear public articulation of the principles of sustainability actually came in the 1980 publication of the World Conservation Strategy.[3] The World Conservation Strategy was developed and written by the International Union for the Conservation of Nature and Natural Resources (or IUCN). The World Conservation Strategy focused on the ecological management of global croplands and watershed forests, and emphasized the need to preserve genetic diversity in these ecosystems and to develop sustainable utilization strategies (IUCN 1980; Adams 1990b: 42–51). While only focusing on the fields of nature conservation and environmental management, the World Conservation Strategy was important to the evolution of sustainability because it made the radical assertion that economic development was not only mutually compatible with nature conservation, but that certain forms of economic development were actually necessary if conservation was going to be achieved. In this context the World Conservation Strategy argued that it was only when tied in with strategies for economic development (whether they be agricultural production, nature tourism, or genetic knowledge transfer) that strategies for nature conservation could be effectively funded and internalized within the everyday practices of those working in delicate ecosystem areas.

At the same time it was argued that only when nature and the environment were properly managed and conserved could lucrative economic practices be realized. This vision of a *positive-sum game* existing between economic development and environmental conservation is characteristic of the ideologies of sustainable development. Sustainable development is the dominant international ideology of sustainability, but as we see later in this book, it is by no means the only version of sustainability in existence. What notions of sustainable development suggest is that we need to be aware not only of the ways in which social, economic and environmental systems inter-react, but that a positive relationship can be established between these fields of existence, whereby environmental conservation can enable economic development, which in turn can alleviate social injustice and suffering, which in turn can contribute to greater efforts in the field of environmental management. While the World Conservation Strategy was heavily criticized for not recognizing the economic and political processes which had created unsustainable practices in sensitive environmental areas in the first place (see Redclift 1987: 21), through the support it gained from the United Nations Environment Programme and World Wildlife Fund, it was able to popularise the notion of sustainable development to a world-wide audience for the first time. From a geographical perspective what is most interesting about the World Conservation Strategy is the way in which it translated local resource management strategies and applied the principles of sustainable yield at a much larger geographical scale.

Philosophical roots of sustainability: green thinking

In addition to its scientific/conservationist origins, it is clear that the notion of sustainability also has a legacy within the intellectual history and philo-sophical development associated with the green movement (see Dobson 1995; Pepper 1996). From the inception of a distinctively anti-industrial set of *green thinkers* in the late eighteenth and early nineteenth centuries in the West, up to the green movement of the mid- to late twentieth century, green ideologies and philosophies have continued to influence the emer-ging concept of sustainability. At the centre of green thinking and phil-osophy has been a continual desire to convey three key messages: (1) that unfettered industrial development can do untold long-term damage to the environment; (2) that contrary to popular industrial ideologies, humanity is still deeply dependent on the environment for its well-being and survival; and (3) that society needs a new system of values that does not simply see the environment as an economic resource, but as something of more varied spiritual, aesthetic and cultural value (for more on the philosophies of the green/environmental movement see Carter 2001: chapter 1). While it is clear that the idea of a sustainable society is something altogether different to the types of society envisaged within what is sometimes referred to as the *deep-green movement* – a movement of ecocentric philosophies inspired by famous green thinkers such as Arne Naess – it does have much in common with so-called lighter shades of green thought (see Box 1.1) (for a detailed review of the different shades of green philosophy see O'Riordan 1989). Consequently, while deep-green thinkers have consist-ently emphasized the need for a kind of *biospherical egalitarianism*, within which the rights of the non-human world (including animals and the environment) are protected through an anti-development agenda of social lifestyle sacrifices, *eco-socialists* and *social ecologists* have focused more on establishing new systems of socio-economic as well as environmental values – something which is much more akin to sustainability.

A note on boxes

The boxes that appear throughout this book have two main purposes. First, certain boxes provide additional background information on people and organizations that are referred to in the text. Second, boxes are also used to describe short cases studies, which serve to illustrate the ideas and principles outlined in the main text. Whether used in the first or second way these boxes are designed to offer additional aids and support to the reader. Every box has been designed as a self-contained entity and as such does not have to be read in order for the chapter in which it appears to make sense.

Box 1.1 Arne Naess, deep ecology and the green critique of sustainable development

Arne Naess is a Norwegian philosopher who is famous for developing the concept of deep ecology. Naess argued that the environmental movement could be split into two distinct categories: the shallow-ecology movement (concerned primarily with the links between human well-being and limited environmental conservation), and deep ecologists (who are more concerned with the intrinsic rights of nature beyond its co-option within human systems). As an advocate a deep ecology, Arne Naess attacked the World Conservation Strategy (1980) for being too anthropocentric (human-focused) and for only valuing the environment when it can bring social and economic benefits. The argument that the World Conservation Strategy – and by extension the principles of sustainable development it proposed – is too anthropocentric has been a common criticism of sustainability which has been used by the deep-green movement. They argue that sustainable development is a set of *technocentric* principles (i.e. human-centred practices and values), which fail to incorporate more *ecocentric* views (i.e. views which prioritize ecological needs and the *ways of nature*) within environmental management.

Key reading: Naess, A. (1994) 'The shallow and deep, long-range ecological movement', in L. Pojman (ed.) *Environmental Ethics: Readings in Theory and Application*; Palmer, J.A. (ed.) (2001) *Fifty Key Thinkers on the Environment*: 213; Adams, W.M. (2001) *Green Development: Environment and Sustainability in the Third World*: 164–167.

If we take a closer look at the writings of prominent eco-socialists and theorists of social ecology, we can see many parallels with notions of sustainability. Inspired by the philosophies of anarchism, Marxism and ecologism, writers such as Rudolf Bahro, E.F Schumacher and Murray Bookchin, for example, all developed community-based visions of environmental reform in the post-war period, which have much in common with certain visions of the sustainable society (see Boxes 1.2–1.4). Focusing primarily on the need for a reconfigured set of relations between the social and natural world, writers like Bahro, Schumacher and Bookchin all argued for the creation of political and economic spaces within which the needs of nature, society and the economy could all be met in a sustainable way.

The emphasis which philosophies of the sustainable society place on socio-environmental change lead many to see sustainability as simply a product of the environmental movement. What I hope that this brief introduction to some of the philosophical origins of sustainability has shown is

---

Box 1.2 Rudolf Bahro and red–green alliances

While working with the West German environmental movement (*Die Grünen*) during the 1970s Bahro wrote about the need to synthesize socialist and green philosophies (see Bahro 1984, 1986). Bahro emphasized that while the environmental movement brought much needed attention to the exploitation of nature (in both the socialist and capitalist worlds), it did not address the important issues of social injustice which were uppermost in Marxist thought. In a world which contained great social injustice (both in relation to class divisions and the great economic disparities between the developed and developing world), Bahro claimed that only a politics which fused environmental and socialist philosophy (a *red–green alliance*) could effectively address the problems of world society. In order to achieve this type of political fusion, Bahro advocated that the green movement should create *liberated spaces*, where people could opt out of industrial society and explore ways in which to achieve more sustainable, socio-ecological patterns of existence. The ideas of Bahro and the wider European eco-socialist movement he animated, are a clear precursor of the types of discussions over the need to achieve social and ecological justice which have become a central part of the sustainability debate.

Key reading: Barry, J. (2001) 'Rudolf Bahro 1935–97', in J.A. Palmer (ed.) *Fifty Key Thinkers on the Environment*: 269–273.

Box 1.3 The small and beautiful world of E.F. Schumacher

In his now famous book *Small is Beautiful* (1973) E.F. Schumacher argued that large-scale social and economic forms of organization were problematic for too reasons: (1) the inherent inefficiency of large-scale socio-economic processes tended to produce waste and to contribute to unnecessary forms of environmental pollution; and (2) the institutional form of large economic operations often alienated workers from the environmental resources (whether it be soil or minerals) upon which their economic activities depended. The environmental results of economic *giantism* were unnecessary environmental pollution and a population with little sense of attachment to, or knowledge of, the environment upon which they depended. In response to the problems of large-scale economic practices, Schumacher proposed the restructuring of economic life around much smaller production complexes. According to Schumacher such smaller-scale units could be highly productive, but they also enabled the purpose of the economy to be altered from being a system of wealth production to a means of producing the types of environment and goods a community needs. The subtitle to Schumacher's book *Small is Beautiful* is *A Study of Economics as if People Mattered*, and it was the reconstruction of the role of the economy to better serve the everyday needs of people which lay at the centre of Schumacher's project. Schumacher's vision of a re-scaled economic system, put to the service of a broad set of social and environmental aims, is an underlying principle of sustainability. The philosophies of sustainability, as with Schumacher's, recognize that while the abandoning of economic growth is not a sustainable environmental or social option, restructuring how and why economic activity occurs should be a pivotal consideration for future social planning.

Key reading: Schumacher, E.F. (1973) *Small is Beautiful: A Study of Economics as if People Mattered*.

that the idea of a sustainable society cannot simply be traced back to environmental thinking. In fact, in many ways the very notion of sustainability (emphasizing as it does the balance of social, economic and environmental considerations) involves a rejection of certain factions of the deep-ecology arm of environmentalism. If sustainability has philosophical roots within environmental thinking, then these roots lie in the types of integrated socio-economic and socio-ecological writings of people like

Box 1.4 Murray Bookchin and social ecology

Writing in a similar vein, and at a similar time, to both Bahro and Schumacher was the influential Russian-American philosopher Murray Bookchin. Bookchin is perhaps best known for his vitriolic attacks on the deep-green movement and his notion of social ecology (Bookchin 1980). In various writings, Bookchin claimed that deep ecology represented a form of environmental extremism which dangerously overemphasized the needs of the natural world above those of humanity (it was not that Bookchin was anti-ecological; indeed he spent most of his life championing green causes and proposing radical green philosophy). In response to the principles of deep ecology, Bookchin proposed his philosophy of social ecology (Bookchin 1986). According to Bookchin the idea of social ecology was essentially a call to recognize the necessary interdependencies which connect the social and ecological worlds, and to develop new eco-communities and eco-technologies which instigated more sustainable relations between these two deeply interrelated worlds. Echoing the anarchist sentiments of Schumacher, Bookchin argued that his vision of social ecology could best be achieved within a set of devolved (or small-scale) communities which were able to exist in greater harmony with the natural world. Unlike the types of organic or biologically determined communities envisaged by many deep-green anarchists, the municipal communities described by Bookchin were designated on the basis of social politics and identity, but were nevertheless geared towards more harmonious modes of socio-ecological existence (Bookchin 1991).

Key reading: Bookchin, M. (1980) *Towards an Ecological Society.*

Bahro, Schumacher and Bookchin. Writing during the 1970s and 1980s, which were the formative years of sustainability, it is clear that such holistic socio-environmental thinking impacted upon the emerging political construction of the sustainable society. What is particularly interesting about these philosophical visions of sustainability is the particular geographical forms they took. What possibly unites Bahro, Schumacher and Bookchin is not so much their search for a sustainable paradigm for future social development, but their consciousness of the new geographical dimensions this society should take. Consequently, Bahro talks of *liberated spaces* of sustainability; Schumacher describes the need for localized economic production and consumption systems; while Bookchin calls for new types of political government, which rather than being based at the nation-state level are delivered through local libertarian municipal associations. As David Harvey (2000) observed, developing new socio-ecological utopias is an inherently geographical practice, and so it seems to be with the sustainable society.

Geopolitical origins of sustainable development

If we need to be aware of developments in ecological science and environmental philosophy to understand how notions of sustainability first emerged, it is only through a study of late twentieth-century geopolitics that we can really appreciate why sustainability has become such an important principle (see Dalby 1996). When I refer to geopolitics here, I am specifically drawing attention to the operation of world politics and the political geography of global space (see Dodds 2000). While geopolitical concerns in the West have historically been focused on issues of territorial expansion (particularly through empire building and imperialism) and military security (for example the bipolar strategic struggles of the Cold War), during the 1960s and early 1970s Western geopolitical attention gradually focused upon a new security issue, the threat of an environmental crisis. Increasingly, evidence from the scientific community suggested that the long wave of industrialization which had been experienced in the Northern hemisphere was creating serious environmental problems, which could, if unchecked, have serious socio-economic consequences.[4] In the light of these warnings, many governments in the West were determined to take action to address the threats to the global environment. However, they found that the global, trans-boundary form of environmental problems made unilateral state action pointless – if the threats to the global environment were to be effectively tackled, an international coalition of states would have to act together in order to address these problems. It is in this context that the notion of sustainable development emerged as a product of an inter-scalar conflict between the interests of individual nation states and the broader global community.

The first attempts to create what Adams (1990b: 32) has described as an *international environmentalism* came in Paris in 1968 with the Intergovernmental Conference of Experts on the Scientific Basis for Rational Use and Conservation of the Biosphere. This conference provided a context within which countries from all around the world could be engaged in discussing issues of environmental conservation and ecological management. The Paris conference was, however, only a precursor to the larger United Nations Conference on the Human Environment (UNCHE) held in Stockholm in 1972 (see Rowland 1973). Following preparatory meetings in the Swiss town of Founex in 1971, the UNCHE brought together representatives from 113 states and over 500 non-governmental organizations.

In all the conference produced twenty-six principles and 109 recommendations for international action on various social and environmental issues. While the UNCHE was heralded as a great success at the time, it was subject to a number of serious limitations. First, following the exclusion of East Germany from the conference, much of the communist bloc (including the Soviet Union) boycotted the event. The failure to command international support from the communist world was compounded by the reservations that the less economically developed countries (LEDCs) attending the conference had. LEDCs attending Stockholm were fearful that the more economically developed countries (MEDCs) were threatening their right to economic development in order to protect the environment. This threat to future economic development was significant, because if the notion of a global environmental crisis was at the forefront of MEDCs' thinking, then alleviating social poverty, deprivation and the conflicts which these conditions produce were the key geopolitical issues facing LEDCs at the time (see Adams 1990b: 36–41). While the Founex and Stockholm conferences suggested that economic development and environmental protection were not mutually exclusive, there was little in the way of any concrete commitment to, or clear articulation of, this principle.

In the aftermath of the Stockholm conference it became clear that a more durable and clearly established set of principles connecting international development with global environment conservation had to be established if effective multilateral action on global socio-ecological problems was to be achieved. It was not, however, until 1983 that the philosophies of sustainability really started entering the mainstream of international politics as a potential solution to competing global political goals. It was in 1983 that the United Nations Secretary General invited the then Prime Minister of Norway, Gro Harlem Brundtland, to establish the World Commission on Environment and Development (WCED). The Commission, which was comprised of people drawn from twenty-two different countries, had the task of developing an agenda for global change which recognized the different social, economic and environmental problems of the world. Drawing on a range of different empirical studies, consultations and discussions, the Commission published its final report, *Our Common Future*, in 1987. Its core recommendation was that the future of the planet depended on developing and implementing the principles of sustainable development. While the WCED's notion of sustainable development was undoubtedly drawn in part from the World Conservation Strategy, *Our Common Future* provided a much more detailed account of what the principles of sustainable development were (see WCED 1987: chapter 2). The WCED also extended the concept of sustainable development from the narrow conservationist concerns of the World Conservation Strategy to cover a bewildering array of issues including: population growth, food supply, species preservation, energy consumption, industrial restructuring, urban development, the global commons, military activity and political institutions.

The doctrine of sustainable development advanced by the WCED defined sustainability in three important ways. First it understood sustainability in relation to inter-generational justice – or that our present economic and environmental practices should take into account the needs of unborn generations to come. Second, it not only recognized that certain forms of social, economic and environmental development were compatible, it suggested that they were necessary prerequisites of each other. The relationship between economic development and environmental protection was perhaps most clearly articulated in *Our Common Future* when it described how social poverty, and the desperate measures it can necessitate, can be more damaging to the environment than carefully practised economic growth and wealth creation. Thirdly, and perhaps most importantly, the WCED illustrated how the principles of sustainability were relevant to all aspects of our lives. No longer could sustainability be consigned to the field of nature conservation – sustainability was about securing global economic, social and environmental security.

By suggesting that it was not economic development which was compromising environmental sustainability, but poverty, and the wrong types of economic practice, it became much easier for the United Nations to engage with LEDCs on issues of sustainable development. Indeed, following the publication of the WCED's report, the second United Nations conference on environment and development was organized and held in Rio de Janeiro in 1992. The purpose of this conference was to try to forge international agreement on the principles of sustainable development. The Rio Earth Summit was officially called the United Nations Conference on Environment and Development (UNCED for short). With the greater sense of international consensus which had been forged around the concept of sustainable development (and the end of the Cold War), the summit attracted representatives from 175 countries and over 1,500 non-governmental organizations. While I do not want to go into detail here, the results of the summit were international agreements on climate change, biodiversity conservation and forest management and a comprehensive vision of sustainable development in the twenty-first century, called Agenda 21 (see Box 1.5). Numerous critiques of the Rio Earth Summit have been written (see Adams 2001: chapter 4; Chatterjee and Finger 1994; Holmberg 1993; Robinson 1993) and I do not want to rehearse their arguments here (though for a summary see Box 1.5). The key consequence of the Rio Earth Summit was that after it, the principles of sustainable development become engrained within international and national policy agendas – the foundations for beginning to build a more sustainable society had seemingly been laid.

Box 1.5 The Rio Earth Summit

The main outputs of the Rio Earth Summit (UNCED) were:

- *The Rio Declaration* (containing 27 principles).
- *Agenda 21* (containing 40 chapters and running to over 600 pages).
- *A Forest Conservation Strategy.*
- *An International Treaty on Biodiversity.*
- *A Climate Change Convention.*

Despite these apparent achievements, the Rio Earth Summit has been criticized from a wide range of perspectives. Many have drawn attention to the failure of the conference to make enough progress on the issue of climate change. Others have claimed that while the Earth Summit recognized the link between environmental destruction and social poverty, it did little to address the trade imbalances and pressures of debt repayments which still blight many LEDCs.

Key reading: Adams, W.M. (2001) *Green Development: Environment and Sustainability in the Third World:* chapter 4.

The latest episode in the historical emergence of sustainable development was played out in Johannesburg in 2002. Ten years on from the Rio Earth Summit, the United Nations convened its third major international conference on issues of the environment and development – this time the name of the conference, the United Nations World Summit on Sustainable Development, expresses clearly the new-found prominence of sustainability within the international political lexicon. The primary purpose of this third world summit was to review the progress which had been made since the Rio conference and to try and develop strategies through which the goals of sustainable development could be more effectively implemented. Approximately 22,000 people attended the summit and most attention was given to the long-standing problems which surrounded international agreements on climate change, the development of renewable energy technologies and the supply of clean drinking water throughout the world.[5] Arguably the most important agreement reached at the Johannesburg summit was the pledge made by world governments to halve the number of people who currently do not have access to clean water or sanitation by 2015 (United Nations 2003a). Far less progress was made on the recalcitrant issues of climate change and renewable energy targets.

From the sustainable society to sustainable societies: towards a geographical perspective

Some people see the history of sustainability I have just outlined as a three-fold process: the first phase involving the conceiving of the notion of sustainability; the second phase concerned with agreeing on the notion of sustainability; and the third phase (the stage we are supposedly now in) being the implementation of sustainability. The problem with this vision of the history of sustainability is that it suggests that the evolution of sustainability has been a process of completion and closure around the concept of sustainable development. But, as we have seen, the history of sustainability is a history of contestation and contradiction, and those conflicts continue today. While sustainable development may be the dominant international vision of socio-ecological development, other more radical visions and interpretations of sustainability coexist with sustainable development (see McManus 1996). In this sense, as a reader you will be consistently encouraged to think of sustainability not so much as an absolute condition, but rather as a socially constructed category. Of course, as soon as you start to think of sustainability as something which is being constantly constructed and reconstructed important questions emerge concerning who is doing the constructing and why certain constructions of sustainability are being promoted and not others (for a wonderful discussion of theories of social construction see Hacking 1999).

Throughout this book we see how thinking geographically helps to reveal the constructed nature of sustainability and uncover alternative sustainabilities, or *countercurrents* to sustainable development (see Adams 2001: chapter 6). But what does thinking geographically actually involve? As we have already established, at one level thinking geographically involves looking at things spatially. Of course when you look at things spatially you immediately become aware of the ways in which interpretations of sustainability vary greatly from place to place – from the developed to the developing world; from rural communities to metropolitan neighbourhoods; from the tropics to the Arctic. Spatial awareness of this kind also indicates the ways in which sustainability is expressed and lived through both the formal spaces of international governments and the state (including planning regions, state departments and local authorities), as well as more radical and liberated spaces (including anarchist communes, local environmental collectives and bioregions). Throughout this book the differences between the formal and liberated spaces of sustainability is a theme to which we continually return.

Of course thinking geographically also has other implications. From the inception of the discipline of geography, geographers have consistently striven to produce integrated social and environmental accounts of the

Earth. In many ways the discipline of geography represents an attempt to institutionally integrate the social and environmental sciences. Despite the growing division which is constructed between human and physical geography, thinking geographically implies thinking about the connections between the social and environmental worlds, about the links between economics and ecology, politics and nature. On these terms, geography is a branch of study which has always sought to think in sustainable terms. The final aspect of thinking geographically is an active concern with issues of scale. Scale is not understood here simply in terms of relative size (something like the region being bigger or smaller than something else – like the city or state), but as a relational category. By relational category, I mean the geographical study of scale reminds us of the ways in which global processes are intertwined with processes operating in states, regions, cities, communities and homes. In the context of the new scales of decision making and action which are becoming important within the sustainable and global society, the geographical concern with scale is a critical component of a *geographical perspective* on the sustainability.

Summary: the geographical exploration of the sustainable society begins . . .

This chapter has illustrated why geography is important not only to the study of the sustainable society, but also to developing an appreciation of the multiple versions of the sustainable society which are emerging in the world today. In the first section we saw how geography's concern with the evolving spaces and scales of globalization mean that it is sensitive to the new spaces and scale of socio-economic life which are prioritized within discussions of sustainability. In the second section we discussed how the desire to create a more sustainable society is partly being driven by fears over the future. These fears have been manifest in a range of dystopic visions of the world. The focus of these dystopic visions on issues of environmental, social and economic crisis also revealed the type of issues which visions of the sustainable society must deal with. The final section of this chapter charted the philosophical and geopolitical origins of sustainability. This section illustrated the ways in which the utopian ideal of the sustainable society incorporates a vision of the world in social, economic and environmental balance, a world in which social and environmental justice are both addressed.

Having established the broad parameters of what I consider to be a geographical perspective on the sustainable society, the remainder of this book analyses various facets of sustainability as they have emerged within specific geographical locations and around particular geographical scales.

Part I of this book considers the emergence of sustainability within different geographical contexts. Focusing on the operationalization of sustainability in more economically developed countries, post-socialist states and less economically developed countries, this part reveals that the application of sustainable development has been a highly contested and complicated process. In this context rather than witnessing the uniform application of a universal set of principles, the geographical spread of sustainability has seen a range of different sustainabilities emerging, which all reflect very different socio-economic priorities and cultural values. Part II of this volume explores the different geographical scales which are being utilised within the development and implementation of a more sustainable society. In the context of the apparent inability of nation states to deal with the complex social, economic and environmental processes which determine sustainability, this section considers the emergence of global strategies for sustainable development and the rise of sustainable regions, cities and citizens. Throughout this section we see not only how these new scales of socio-political organization are offering new ways of approaching questions of sustainability, but how they are also providing new ways of understanding and restructuring the spatial relations of human society. Through its two-part structure it is hoped that this volume will mirror the types of geographical ways of thinking about sustainability which it is trying to promote.

Suggested reading

There is actually very little which currently links geographical study with the analysis of the sustainable society in any explicit way. This said a good starting point for understanding the value of a geographical perspective on the world is provided by: Adams, W.M. (2001) *Green Development: Environment and Sustainability in the Third World* (in the context of the Third World); and Cloke, P., Crang, P. and Goodwin, M. (eds) (1999) *Introducing Human Geographies* (section 3) (for a more general introduction to how geographers approach issues of sustainability). A more difficult, but nevertheless rewarding, description of the role of spatial analysis within the social sciences is provided by Soja, E. (1989) *Postmodern Geographies: The Reinsertion of Space in Critical Social Science*.
In terms of work on the historical emergence of sustainable development the books by Dresner, S. (2002) *The Principles of Sustainability* and Adams, W.M. (2001) *Green Development: Environment and Sustainability in the Third World* are unsurpassed in the quality and depth of their coverage.
Finally for those who may be interested in the links between sustainability and environmental philosophy I can recommend Dobson, A. (1995) *Green Political Thought* and Pepper, D. (1996) *Modern Environmentalism: An Introduction*. For a more detailed look at the thoughts and writings of different environmental philosophers I can also recommend Palmer, J.A.

(ed.) (2001) *Fifty Key Thinkers on the Environment* (which includes sections on Gro Harlem Brundtland, E.F. Schumacher, Rudolf Bahro, Murray Bookchin and Arne Naess).

Suggested websites

Too many web resources exist outlining the principles of sustainability and the politics of sustainable development to detail here. One site I would however recommend is the United Nations Environment Programme's own website: http://www.unep.org/. Links can be found through the 'Milestones' section of this page to sites devoted to all of the major UN conferences on environment and development (from Stockholm to Johannesburg).

PART I
Spaces of sustainability

2 ECOLOGICAL MODERNIZATION IN THE WEST

Making business sense out of sustainability

Introduction

In the previous chapter we discovered how concerns over the global environment in more economically developed countries (hereafter MEDCs) played a crucial role in the emergence of sustainable development as a geopolitical strategy, for socio-economic and environmental development, at an international level. Despite these strong historical links between the geopolitical priorities of MEDCs and sustainability, two interesting processes have been emerging in MEDCs with regard to sustainable development policies. First, MEDCs have been gradually developing their own particular brand of business-friendly sustainability, which is customarily referred to as *ecological modernization*. Second, in certain MEDCs, we are seeing the beginnings of a backlash against the principles of sustainable development (Rowell 1996). Such a backlash, which actually questions the validity and value of certain principles associated with sustainability, has led some to believe that just as the path towards a sustainable society began in the MEDCs, it may already be reaching the end of its life there.

This chapter explores the evolving politics of sustainability in MEDCs through four interrelated sections. The first section explores how and why sustainability in MEDCs has increasingly taken the form of business-oriented practices and discourses of ecological modernization. The second section then focuses upon the United Kingdom and the ways in which the ideologies of ecological modernization have purportedly played a crucial role in Britain's transformation from 'dirty man of Europe' to an international leader in the development and implementation of socio-environmental policy. In the third section, we see how in the USA the economic logics of ecological modernization and sustainability are being questioned and challenged by the business community with often disturbing consequences for environmental policymakers. In light of the reactions to ecological modernization and sustainability in countries like Britain and the USA, the final section of this chapter explores some alternative spaces within which new visions of sustainability are being forged in the MEDCs. I call these alternative spaces, sustainable hetero-topias, because they present multiple and often overlooked visions of a sustainable society which have been inspired by radical green thinking or anarchist philosophies.

Exploring ecological modernization: making business sense out of sustainability

Rarely it seems can I escape a family shopping trip without a now routine visit to the Body Shop. No matter where we may go, or which city we may be in, I can't seem to avoid Body Shops. Perhaps, I shouldn't find this situation too surprising given that the Body Shop plc currently has stores in fifty different countries and has over 1,900 outlets, which operate using twenty-five different languages across twelve different time zones – there is no escape! The reason I mention the Body Shop is, however, not merely to exercise my own consumer-based suffering, but because my numerous visits to this chain store have led me to realize that the Body Shop represents a quintessential expression of the sustainable society. A visit to the company's Australian home page helps to illustrate my argu-ment (go to http://www.thebodyshop.com.au/). The thing about this home page is that while the page enables you to join the Body Shop's multiple campaigns to protect the planet, defend human rights, stop animal testing, establish fair trade agreements or stop household violence, and helps you find out more about the values of the Body Shop and how to get involved in various fund-raising exercises, there is something missing – the company's products. When you do finally find associated web pages which tell you about the Body Shop's products you discover an array of exotic sounding items including Brazil Nut Body Butter, African Spa Rich Cocoa Body Balm and Grapeseed Glycerin Soap. Accompanying each of these products are explanations of how and why these items are produced in ethically sound ways, through fair trade agreements and have benign environmental effects.

Despite its lack of apparent emphasis on the products its sells, the Body Shop represents the ways in which the business community have made economic sense from sustainability. The reason why the Body Shop seems to pay such little attention to marketing its products is because it is not really functional goods which it is trying to sell – I mean do you really buy Grapeseed Glycerin Soap just because it makes you clean? What the Body Shop does sell is a brand, or, put another way, a set of mean-ings (see Klein 2001: chapter 1 for more on theories of product branding). Consequently, to purchase items from the Body Shop is to buy in to an ethically sound, sustainable lifestyle first and a bar of soap second. Listen to what Naomi Klein, a commentator on our branded lifestyles, has to say about the Body Shop:

> The Body Shop had been present in Britain since the seventies, but it wasn't until 1988 that it began sprouting like a green weed on every street corner in the U.S. Even during the darkest years of the recession, the company opened between forty and fifty American stores a year. Most baffling of all to Wall Street, it pulled off the expansion without spending a dime on advertising. Who needs billboards and magazine adds when retail outlets were three-dimensional advertisements for an ethical and ecological approach to cosmetics? The Body Shop was all brand.
>
> (Klein 2001: 20)

While in the past the idea of protecting the environment, or ensuring the welfare of labourers, was interpreted as a drain on economic development, as Klein indicates, to be seen to be sustainable in today's world – as the Body Shop has historically prided itself on being – can bring with it great economic benefits. By aligning itself with social and environmental sustainability, the Body Shop's success is perhaps best expressed by the fact that it is now estimated that the Body Shop sells one of its ethical products every 0.4 seconds.[1] As Doyle and McEachern (1998: 136–137) point out, the Body Shop's economic success is not just based upon its ethical credentials, but on the apparent socio-environmental failings of other companies. This has led to competitors trying to undermine the ethical claims of the Body Shop and to an emerging ethical consumption battle on the high street (ibid.).

In this chapter we see how the economic success of the Body Shop is indicative of a broader ideological shift within Westernized market economies.[2] Some have described this transformation as the *greening of business*, others as *environmental marketing* (see Figure 2.1), but I believe that only the notion of *ecological modernization* fully captures the extent and nature of this economic movement. It is important at this stage to note that the activities and policies of the Body Shop are only illustrative of a small part of the broader processes which are associated with ecological modernization. The term ecological modernization was first used in the early 1980s by the German political sociologists Joseph Huber and Martin Janicke. While it remains a predominantly European concept, it is clear that the underlying principles of ecological modernization have affected sustainable development policies in many other states and communities throughout the world (for an analysis of the spread of ecological modernization into the Asia-Pacific region for example see Welford and Hills

Figure 2.1 Green marketing and ecological modernization (advert by British Petroleum)

2003). The fundamental premise of ecological modernization is a desire to reconcile the seemingly competing interests of economic development and environmental protection (see Huber 2000; Mol 1999). Echoing the case we have discussed of the Body Shop, the starting principle of ecological modernization is that environmental care is good for business. While the example of the Body Shop illustrates that the promotion of ethical goods and a green image can be good for a retail outlet, this chapter illustrates how ecological modernization has also underscored broader forms of economic change within manufacturing communities. Dryzek and Schlosberg (2001) provide us with a useful way of understanding the range of benefits classically associated with ecological modernization:

> The essential idea [of ecological modernization] is that a clean environment is actually good for business, for it connotes happy and healthy workers, profits for companies developing conservation technologies or selling green products, high quality material inputs in to production (e.g. clean air and water), and efficiency in materials usage. Pollution, on the other hand, indicates wasteful use of materials [. . .] It is cheaper to tackle environmental problems before they get out of hand and require expensive remedial action.
>
> (Dryzek and Schlosberg 2001: 299)

This quote serves to emphasize the *double logic* of ecological modernization. This double logic is premised upon the realisation that not only can sound environmental policies save money and generate added revenue, but that a poor environmental record can be costly for a company (or nation state).

In essence ecological modernization reaffirms the virtuous link established between economic growth and environmental protection within contemporary notions of sustainability (see Huber 2000). In this sense, ecological modernization asserts that it is not economic growth or development which is harmful to the environment, but the 'wrong type' of economic growth or development. It is in this context that Huber (2000) contrasts the *sufficiency* discourses of green-oriented versions of sustainability (suggesting a limit to necessary economic growth once sufficiency has been reached), and the *efficiency* discourses of ecological modernization (intimating the importance of continued economic growth based on the more efficient use of environmental resources). Of course modernization has been a key characteristic of capitalist economics for the last two hundred years. Conventionally, modernization has been premised on the strengthening of a company's market position, either through new product innovations, or by developing more efficient production technologies and working practices. Significantly, not only were these historical innovations often not concerned with issues of environmental protection, routinely they were based upon regimes of increased environmental exploitation and ecological pollution. Ecological modernization reasserts the importance of economic modernization, but claims that now modernization should not simply be based upon economic calculations and technological change, but also on ecological considerations. In this context, the idea of ecological modernization is perhaps best thought of as an economic/environmental cycle, within which economic growth and environmental protection not only coexist, but form a system of mutual support (see Figure 2.2).

The specific link which ecological modernization forges between environmental protection and increased economic profits is important for two reasons. First, because it challenges the conventional wisdom that the maintenance of economic profit margins requires the continued exploitation of *environmental externalities* (or those things in the environment which are used by businesses cost-free – i.e. air and water). Second, it emphasizes the direct benefits of sustainable business policies to the individual industrialist or businessperson. In this sense, ecological modernization takes the laudable but general principles of sustainability and relates them directly to the workings of the business community. This is important, because while most in the business community would support the principles of establishing a sustainable social, economic and environmental future (after all the continuation of their business could well be

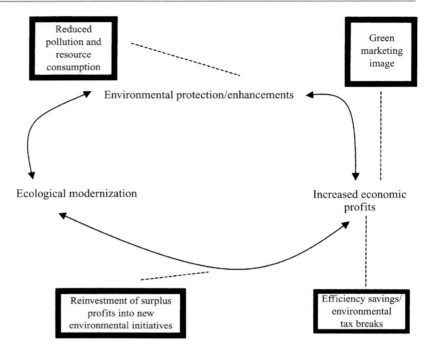

Figure 2.2 The cycle of ecological modernization

compromised in an unsustainable future world), few would be willing to endanger their profit margins on the basis of socio-environmental benevolence alone.

Although offering the opportunity to foster the engagement of the business community within the formation of a more sustainable society, some writers have been critical of the fundamental assumptions of ecological modernization. The basis for these criticisms has been a belief that in gaining the support of certain factions of the business community for sustainable development, some of the most important principles of sustainability may have been compromised. Perhaps the most influential writer on, and critic of, ecological modernization, has been Maarten Hajer (see Hajer 1997; Hajer and Fisher 1999). Essentially Hajer argues that as it is currently constructed, ecological modernization transforms sustainability into something that maintains modern techno-industrial practices and which can be easily assimilated into a relatively unchanged business community. In this context, Hajer states that:

> *ecological modernisation* can be defined as the discourse that recognises the structural character of the environmental problematic but nonetheless assumes that existing political, economic and

social institutions can internalise the care of the environment. For this purpose ecological modernisation, first and foremost, introduces concepts that make issues of environmental degradation calculable.

(Hajer 1997: 25–26)

Hajer is critical of ecological modernization on two fronts: (1) it is overly conformist – that is to say that it does not call for fundamental changes in how the modern economy works; and (2) it tends to reduce notions of environmental value to economic profit. Crucially, the work of Hajer questions the assumption which ecological modernization makes that our current economic system can be made to be sustainable.

Beyond the concerns raised by academics over the long-term veracity of ecological modernization, significant anxieties have also been raised within the business community itself. Many, for example, feel that the principles of ecological modernization are more easily applied to certain sectors of the economy than others (for example in the service sector as opposed to mining), while being more easily implemented within the relative economic stability provided by affluent states as opposed to more impoverished social situations (see Doyle and McEachern 1998: chapter 6). It is also becoming increasingly apparent that companies are only sustaining the apparent win–win scenario which ecological modernization appears to offer because they deliberately offset the socio-environmental costs and injustices of their activities to other spatial scales. With the social and environmental relations of corporations becoming increasingly global, it appears that companies are becoming increasingly adept at sustaining an ethical image at a local level, while continuing to cause socio-environmental damage in other, more distant locations and scales (for more on the links between scale and contemporary environmental policies see Bulkeley 2005; Bulkeley and Betsill 2005). The remainder of this chapter explores the impact ecological modernization and reform are having on the economies and politics of the United Kingdom and USA.

Ecological modernization and the UK: improving the national quality of life

The United Kingdom provides a very interesting case study of ecological modernization in action. This interest derives from the fact that despite recently embracing ecological modernization as a key principle for its broader policies for sustainable development, for many years the British government rejected ecological modernization in favour of a much more bureaucratic and sceptical approach to industrial–environmental policy. In this context the transformation of the British policy scene provides us with some interesting insights into the nature of ecological modernization and its competing models of sustainable development policy.

Pollution regulation and end-of-pipe solutions: the 1970s and 1980s

When it comes to discussions of sustainability, the UK is something of a paradox. At one level it can be seen as a world leader – it was the UK's 1956 Clean Air Act which is widely regarded as the first piece of environmental legislation focusing on air pollution passed in the world; while the establishment of the Ecology Party in 1973 in the UK was the first formal green party in Europe. The UK government also took a full and active part in the United Nations Conference on the Human Environment in Stockholm. Yet despite these achievements, the UK has continued to endure a poor record when it comes to social and environmental sustainability. As one of the first industrialized nations in the world, the UK had built up a legacy of severe environmental pollution and poor living conditions for workers in and around heavy industrial districts. Many of Britain's most acute environmental problems were related to air pollution. By the early 1960s, the UK had one of the worst air pollution records in Europe with sulphur dioxide emissions (mainly produced by power stations) standing at 179.6 $\mu g/m^3$ and associated black smoke concentrations of 155.8 $\mu g/m^3$ (Environment Agency 2004). While there were concerns over the local health implications of such pollution concentrations, there was also a growing concern in continental Europe that British air pollution, carried by prevailing winds, was causing serious environmental damage in other countries.

In the context of domestic and international concern over British environmental pollution, in 1970 the British government established a Royal Commission on Environmental Pollution to try and think through how to restructure Britain's industrial relations with the environment (see Owens and Rayner 1999; Owens 2003). The Royal Commission published its first report on the government's environmental pollution abatement strategies in 1971 and provided subsequent reports on industrial air pollution in 1972 and 1973. Despite the early focus on issues of air pollution, the Royal Commission gradually widened its studies of pollution in the UK to incorporate analyses of river pollution (1973), pollution in estuaries and coastal waters (1972), nuclear power (1976), lead contaminates (1983) and pesticide use (1974). Following these reports, the UK government attempted to develop an integrated network of environmental monitoring agencies and Acts of legislation designed to regulate atmospheric, water-based and terrestrial pollution in Britain. This was a system which operated through local authorities, national pollution inspectorates, environment agencies and departments and numerous pollution abatement Acts. This system of environmental regulation was based on two principles: (1) the coercive power of the state to force polluters to adhere to environmental restrictions or face legal or financial penalties; and (2) a related assumption that environmental regulation was essentially bad for business – hence the need for state coercion and power.

Throughout the early period of state-based environmental regulation in Britain, concerns over the effects of environmental restrictions on the international competitive power of British industries meant that the state only tended to act in circumstances whereby it was clear that pollution was causing actual social and/or environmental harm. In this way, the British government was only willing to intervene within economic practices where there was a *burden of scientific proof* to support the validity of state interference (Weale 2001). This model of state intervention differed markedly from that practised by the West German state at the time, where the government adopted a principle of *Vorsorgeprinzip* – a *precautionary principle* (see Weale 2001: 310). So, unlike the British state, the West German government was willing to intervene within economic activity even when there was only a chance that such activity could cause long-term environmental harm. With a reluctance to adopt a precautionary approach to industrial regulation, the British state tended to favour end-of-pipe solutions to environmental pollution, when action was deemed scientifically necessary. The idea behind end-of-pipe solutions and technologies was that rather than stopping the production of pollution, they treat pollution at the end of the production process in order to minimize its environmental effects. End-of-pipe solutions can range from smoke filters fitted on factory chimneys to waste-water pollution treatment plants. Of course the major advantage of end-of-pipe technologies is that while they require some extra cost to the industrial plant or factory, they do not necessitate a more radical and expensive overhaul of the production process itself in order to minimize the very production of pollution in the first place.

Following the emergence of state-based environmental regulation in Britain during the 1970s, the 1980s saw a slight change of political emphasis. With the election of the Thatcher government in 1979, British politics began to be driven by the principles of neo-liberalism. At the centre of Thatcherite neo-liberalism were two key orthodoxies: (1) to reduce and tightly control public spending; and (2) to limit the amount of state intervention within the free-market economy (Mohan 1989). According to the Thatcher government, excessive intervention by the state within national economies tended to burden the entrepreneurial and competitive dynamics of a country's economy. In the context of these twin political and economic beliefs, the 1980s were a paradoxical time with regard to environmental regulation in the UK (see Lowe and Flynn 1989) (see Box 2.1). At one level the Thatcher government sought to reduce the amount of environmental bureaucracy and regulation which British companies experienced. The removal of state regulation, however, did not only affect the private sector. In line with its neo-liberal doctrines, the Thatcher government gradually started selling off large state-owned public utilities like water authorities and electricity generation plants, in order to incorporate free-market practices into the public sector. The practices of public and private deregulation did, however, lead to widespread concern over the welfare of the British environment during the late 1980s. By the end of the decade it became clear neither old-styled, state-centred regulation, nor uncontrolled deregulation, were satisfactory ways of managing industrial relations with the environment in the UK.

Box 2.1 Thatcherism and the environment

Margaret Thatcher had a fluctuating relationship with environmental issues during her tenure as prime minister of the UK. At one level, it was clear that the implementation of state-sponsored environmental policies was very much anathema to her neo-liberal political beliefs. Thatcher was keen to rid Britain of big government and the excessive state intervention which she blamed for economic inefficiency and decline in the UK. In this sense, the idea of environmental controls ran counter to her desire to allow companies greater freedom to generate a healthier economy and a more prosperous nation. Perhaps this type of thinking was the reason why Thatcher once described the environment as a 'hum-drum issue' in the early part of her tenure. Towards the end of the 1980s, however, it became clear that the environment was an important electoral issue and it was probably this which led Thatcher, in her famous Royal Society speech, to claim that Conservatives were guardians of the environment after all. This shift in emphasis was partly made possible by the fact that the Conservatives had historically drawn a great amount of political support from the rural shires and elites. In this context, the defence of the environment become synonymous in Conservative minds with the defence of rural values against the democratic socialism associated with large urban conurbations in the UK.

Key reading: McCormick, J. (1991) *British Politics and the Environment.*

Making business sense out of sustainability: the 1990s onwards

Two key events marked a metamorphosis in British political attitudes to environmental sustainability in the late 1980s. First, in 1988, the prime minister, Margaret Thatcher, signalled a changing set of relations between the Conservative government of the time and environmental policy. In a now famous speech which Thatcher delivered to the Royal Society, she pronounced:

> We Conservatives are not merely friends of the Earth, we are its guardians and trustees for future generations to come. No generation has a freehold on this Earth. All we have is a life tenancy with a full repairing lease.
>
> (Margaret Thatcher 1988, quoted in McCormick 1991: 60)

No longer, it seemed, was environmental protection anathema to the neo-liberal values of the Conservatives, but a guiding principle. Following Margaret Thatcher's Royal Society speech, further pressure was applied to the Thatcher administration to revise its approach to the environment in 1989. In the European elections of 1989, the British Green Party gained 15 per cent of the national vote. This represented the highest proportion of any national vote achieved by a green party in Europe at the time. It was clear that ambivalence to issues of environmental protection and sustainability were no longer a political option in the UK.

In 1990, the Thatcher government published Britain's first environmental strategy – *This Common Inheritance* (HMSO 1990). The principle behind this strategy was to set in place an integrated framework within which government policy on energy, town planning, pollution control and waste disposal could be brought together and harmonized. While providing an integrated perspective on the UK's diffuse environmental policy arenas, the strategy as a whole did not offer a new approach to tackling the tensions between environmental protection and economic security, and continued to support a fairly traditional system of pollution control and monitoring.[3] Two years after the publication of the first environmental strategy in the UK, a transformation occurred in the nature of British environmental policy. Following the Rio Earth Summit of 1992 (see Chapter 1), the UK government agreed to produce a national programme of sustainable development. In creating this new programme – *Sustainable Development: The UK Strategy* (DoE 1994), the British government largely recast its nascent national strategy for environmental protection in terms of sustainability. This was significant, because it provided an opportunity to revise

> **Box 2.2 Key publications and events in the emergence of ecological modernization in the UK**
>
> ▪ *This Common Inheritance: Britain's Environmental Strategy* (HMSO 1990).
> ▪ *Sustainable Development: The UK Strategy* (DoE 1994).
> ▪ Launch of British Governmental Panel on Sustainable Development (1994).
> ▪ Launch of United Kingdom Round Table on Sustainable Development (1994).
> ▪ *A Better Quality of Life: A Strategy for Sustainable Development in the UK* (DETR 1999).
> ▪ Launch of the Sustainable Development Commission (2000).
> ▪ Launch of 'Taking it on: developing a UK sustainable development strategy' consultation process for new national strategy for sustainable development (2004).

the somewhat problematic relationship which the government had created between environmental protection and economic development.

Essentially, the production of the UK strategy for sustainable development enabled the government to explore alternative ways in which environmental reform could be delivered within the British economy. It is clear that from 1994 onwards the UK government began to consider environmental protection not as a barrier, or restriction to economic growth, but as a potential way of enhancing economic performance and efficiency. In this context, the association made between environmental reform in industry and improved economic efficiency was strangely akin to the government's existing neo-liberal doctrines. To support the government's desire to create a more sustainable Britain, the state also created two new institutions in 1994 – the British Governmental Panel on Sustainable Development and the United Kingdom Round Table on Sustainable Development. While often overlooked, these two institutions (which were effectively amalgamated in 2000 with the creation of a new Sustainable Development Commission) were designed to integrate policies for sustainable development across government departments. These two bodies essentially acted as consensus-building think tanks, bringing together key representatives from Britain's industrial and environmental sectors in order to devise mutually acceptable forms of policies for sustainable development. Crucially, what these two institutions sought to initiate in the UK were a set of policies which recognized the potentially beneficial links which could be forged between environmental reform in business and continued (and even improved) economic performance. Despite such laudable goals, however, some have been critical of the real impact these institutions were able to have as they became increasingly marginalized within key policy debates.

Although new policies for ecological modernization in Britain were being developed throughout the 1990s, it was not really until 1997, and the election of the New Labour government, that ecological modernization became a widespread principle of UK policy (see Barry and Paterson 2003). Following their election, New Labour were keen to use the principles of sustainable development to influence an increasingly large spectrum of policy areas – ranging from urban renewal to energy and employment policies. In this context, it was clear that sustainable development provided an expedient political framework within which to develop Labour's new brand of social democracy, which sought to balance the economic rationalities of Thatcherism with a concern for social and environmental justice. The importance of sustainability to the New Labour government was signalled in 1999, when they produced a new national strategy for sustainable development in the UK – *A Better Quality of Life: A Strategy for Sustainable Development in the UK* (DETR 1999). In this strategy the New Labour government set out their vision for social, economic and environmental policy and of how to ensure a *better quality of life* for people living in the UK. Significantly, within this new vision of sustainable development the prime minister, Tony Blair, asserts the important link between continued economic growth and enhanced social and environmental welfare which lies at the heart of ecological modernization thinking:

> we need more not less growth [. . .] We need increased prosperity, so that everyone can share in higher living standards and job opportunities in a fairer society. We must close the gap between productivity and incomes in the UK and those in North America and much of Europe. Abandoning economic growth is not a sustainable development option: to do so would close off opportunities to improve the quality of life through better health care, education, and housing.
>
> (DETR 1999: 1)

Of course, when Tony Blair talks of the need for more, not less, economic growth, he is not simply reasserting the neo-liberal mantra of 'letting the market rip', but is arguing for more of a certain type of economic growth – a type of economic growth which is efficient and more profitable precisely because of its environmental sensibilities.

Following the publication of the UK's new national strategy for sustainable development, the British government has been actively pursuing economic reform through the principles of ecological modernization. The implications of ecological modernization for the British economy was emphasized by the UK's Department for the Environment, Food and Rural Affairs:

> For business, sustainable consumption and production requires consideration of the implications for their business model together with their product and service range. Success will depend on their ability to meet growing consumer (household and supplychain) expectations of higher environmental and ethical standards and to cut out the negative impacts of growing material resource consumption. Businesses that anticipate this trend and develop 'material light' goods and services will be best placed to benefit from these opportunities and to enhance their competitiveness.
>
> (DEFRA 2004: 46)

The desire to modernize Britain's economy using a strategy of greening business has been approached through a number of schemes. These have ranged from supporting a new sustainable production and consumption scheme; helping improve companies' resource efficiency; an integrated product policy to reduce the environmental impact of a product over its whole life cycle; improving consumer knowledge of green practices through a *Shoppers' Guide to Green Labels*; and the introduction of a controversial eco-levy to penalize companies who are inefficient in their energy use. Although the government's eco-levy has proved very unpopular with British manufacturers, who claim it is affecting the international competitiveness of their companies, its does represent a classic example of ecological modernization policy. The idea behind the eco-levy is that British companies are assessed and monitored on their energy use and should they exceed specified levels they are fined. While at one level this strategy may seem reminiscent of traditional forms of state regulation, the eco-levy is designed to highlight the broader cost savings which can be made from more efficient business strategies and technologies. It is therefore significant that the money raised through the eco-levy in the UK is being reinvested into programmes dedicated to the development and use of green technologies. (For more on the new policy instruments for sustainable development being implemented by the New Labour government see Jordan *et al.* 2003.)

One good example of how the British government has sought to use the principles of sustainability, and more specifically ecological modernization, and apply them to the broader economic modernization of the UK, can be discerned in the operations of the Carbon Trust. The New Labour government established the Carbon Trust in March 2001 in order to support the UK's climate change programme and its associated commitment to reduce national carbon dioxide emissions. While supported and partly funded by the government, the Carbon Trust is actually an independent body staffed by scientific and business experts. As an independent body, the government hopes that the Carbon Trust will be able to reach and engage businesses more effectively within its policies for sustainable development. The Carbon Trust has three main missions: (1) to help UK businesses and public sector agencies reduce their greenhouse gas emissions; (2) to promote low-carbon technology research and development; and (3) to help the UK move towards a low-carbon economy. The current strategies which the Carbon Trust are utilizing in order to try and promote a low-carbon economy among British businesses are typical of the types of ecological modernization practices which are now being established throughout Britain.

The Carbon Trust has two main schemes through which it attempts to foster ecological modernization: (1) Action Energy; and (2) Carbon Management. Action Energy, which is based upon experts from the Carbon Trust assessing different companies' everyday energy uses, recommends ways in which energy can be saved or used more efficiently. The Carbon Management scheme on the other hand serves to illustrate to companies the economic benefits which can be gained from reducing carbon emissions – in terms of avoiding carbon levies; making energy efficiency savings; and improving a company's green credentials and associated marketing potential. In this context, the Carbon Trust emphasizes to companies the direct economic savings which environmental reform can yield and the indirect market advantages which becoming a green business can bring. In relation to the marketing potential of green economic reform, the Carbon Trust is also helping businesses to develop advertising strategies which highlight their green credentials. At present the Carbon Trust is working with fifty companies (sixteen of which are in the FTSE 100 – an index of Britain's largest and most successful firms). What is interesting about the links which the Carbon Trust has forged with these companies is that they are based upon companies realizing that it is in their own best (perhaps even selfish) interests to adopt the principles of ecological modernization. The work of the Carbon Trust is also interesting because it reveals the changing role of the state within environmental management associated with ecological modernization. Thus, rather than directly governing the environment through laws and different forms of environmental enforcement, the British state is now increasingly using complex governance networks of NGOs and affiliated bodies (like the Carbon Trust), along with new incentive mechanisms to pursue its sustainable development goals (see Jordan *et al.* 2005).

Although the principles of ecological modernization have become popular throughout many European states, it is clear that they have played a pivotal role in the recent historical emergence of sustainability within British politics. In essence, ecological modernization has enabled an economically acceptable version of sustainable development to flourish in the UK. In turn, this has allowed the UK to move from being the dirty man of Europe, to one of the leading international advocates of sustainability. (Significantly, this shift in environmental policy has also been matched by the UK government taking an increasingly leading role on issues of international poverty alleviation and social justice through its Presidency of the European Union and its hosting of the Gleneagles G8 Summit in 2005.) Perhaps one indicator of the success of ecological modernization in the UK is that Britain was one of only four countries in the European Union (along with Sweden, Germany and Luxembourg) to be ahead of its target for greenhouse gas emissions reduction by 2001 (DEFRA 2003: 10). Many claim, however, that Britain's success in reducing its emissions of greenhouse gases, as well as its discharge of sulphur dioxide and particulate matter into the air, has been a side-effect of the decline of large-scale heavy industry in the UK over the last thirty years and, as such, has relatively little to do with ecological modernization. Recent reports also indicate that British carbon dioxide emissions may be increasing again, and with the Confederation of British Industry strongly opposed to further restrictions on British companies' carbon dioxide allowances, it appears that much work remains to be done. While the precise effects of ecological modernization remain difficult to accurately assess, however, it is clear that it has enabled sustainability to become an acceptable principle even in a country with a long legacy of neo-liberal social, economic and environmental policy.

The sustainable backlash: questioning the economic viability of the sustainable society in the USA

British political enthusiasm for an economically beneficial brand of sustainable development culminated in 2002 at the Johannesburg World Summit on Sustainable Development. At this conference the British government delegation pushed for key carbon emissions reductions and for new global investments in alternative, non-fossil-based energy production. The attitude adopted by the USA's delegation at the Johannesburg conference, however, marked a clear contrast to that of Britain. A clue to the position adopted by the US government can be found in a letter sent to congratulate George W. Bush for not attending the Johannesburg summit:

Even more than the Earth Summit in Rio in 1992, the Johannesburg Summit will provide a global media stage for many of the
most irresponsible and destructive elements involved in critical
economic and environmental issues. Your presence [as the
President of the USA] would only help publicize and make more
credible their various anti-freedom, anti-people, anti-globalisation
and anti-western agendas.

(*The Guardian* 2002: 4)[4]

In part reflecting these views, the US delegation to Johannesburg resisted
calls for it to ratify the Kyoto agreement on climate change and offered
serious opposition to European-backed desires for targets to be set on
renewable energy development. Suddenly, far from being a facilitator
of continued capitalist economic development, to some in the USA, sustainable development had become synonymous with political repression,
anti-humanism and anti-Western agendas. In order to understand the
contemporary political hostility being expressed towards the principles of
sustainable development in certain quarters in the USA (and indeed other
industrial countries), we need to take a brief historical journey into the
socio-environmental politics of the USA (for a critical review of current
US policies for the environment and sustainable development see Pope and
Rauber 2004).

Wattism, the Pacific Legal Foundation and the beginnings of the 'green backlash'

The USA has a long history of forward thinking when it comes to environmental policies and initiatives. From the creation of the first national parks
for the protection of nature, to the establishment of the Environmental
Protection Agency (hereafter EPA) under the Nixon administration in
1970, the USA has often provided an example to other countries in how
to deal with socio-environmental problems. In many ways, however, the
proliferation of environmental initiatives and controls in the USA lies at
the heart of contemporary suspicions surrounding sustainability and
ecological modernization. For example after its inception in the early 1970s
the EPA became a keen enforcer of environmental legislation and took the
fight over environmental quality to corporate America. The first adminis-
trator of the EPA, William Ruckelshaus, worked hard to challenge corporate giants like Union Carbide and the Jones and Laughlin Steel Company
to improve their environmental records (Environmental Protection Agency
1990). While generally perceived to be successful in reforming many of
the environmental practices of the USA's largest industrial corporations,
it is interesting to note that the EPA used fairly traditional methods for

instigating environmental reform. Essentially the EPA utilized existing and new environmental legislation (like the Clean Air Act of 1963 and the Federal Water Pollution Control Act of 1973), violation notices and the threat of legal action to persuade polluting companies to change their ways. In this context, it is perhaps significant that before taking charge at the EPA, William Ruckelshaus was actually an Assistant Attorney General (Lewis 1985).

In the early 1970s, however, industrial corporations in the USA began to co-ordinate a series of legal campaigns to challenge state-based environmental enforcement. An important example of the industrial lobby's campaign was the Pacific Legal Foundation.[5] Founded on 5 March 1973, the Pacific Legal Foundation (hereafter PLF) was established to provide institutional support for those fighting different aspects of government intervention in corporate and private affairs. According to the PLF, big government and excessive legal controls had been strangling the liberties of US citizens and businesses. In this context, the PLF offered legal support to a range of institutions fighting state regulations related to, among other things, issues of environmental protection. The ideologies behind the PLF are perhaps best expressed in this quote which has been taken from their website:[6]

> Liberty In Crisis – Conceived in pain over two hundred years ago, America still stands as 'the land of freedom.' It is still, by far, the greatest country in the world to live in. But there are tarnishes that have formed and need to be removed. Though the great principles of our Founding Fathers remain with us, many of their hard fought protections are eroding due to a form of tyranny engendered by overzealous bureaucracies, red tape, ignorance or indifference of our courts and elected officials, and a complex maze of laws and regulations that are strangling our personal and professional lives.
>
> (Pacific Legal Foundation 2005)

In the purported defence of 'national values', the Pacific Legal Foundation has become actively engaged in a range of high-profile legal cases relating to issues of environmental protection and sustainability. In this context, the PLF's position on environmental legislation is captured in the following statement:

> Americans want a clean, healthy environment. They also want a strong economy and a high overall quality of life. Accordingly, environmental/endangered species protection laws and public land management policies should be measured against the cost to people. Paradoxically, too many of our environmental laws and

regulations, albeit well-intentioned, are unfairly punitive and senseless – i.e., they are based on "junk science," disregard human health and safety, undermine private property rights, impede economic growth and technological innovation, and do little to improve the environment. Pacific Legal Foundation believes that good law should start with people and that good intentions are not enough in developing environmental policy. Thus, PLF challenges in court heavy-handed laws and regulatory actions that endanger human lives, undermine private property, unjustly limit the use of natural resources or impose exorbitant penalties for activities that do not pose a significant, immediate threat to health and safety or to environmental resources.

(Pacific Legal Foundation 2004)

In the context of this hostility towards state-based environmental legislation, the PLF has fought legal battles over the Clean Water Act, natural resource management issues and the controversial Endangered Species Act. Through the association of its message with American values of freedom and endeavour, the Pacific Legal Foundation has managed to gain a broad base of support for its work and agenda.

The popular discontent over the effects of environmental control and regulation in the USA during the 1970s gained more formal political recognition and support during the 1980s. In 1981 the then president Ronald Reagan appointed James G. Watt as Secretary of State for the Interior (this is in part equivalent to the USA's minister for the environment with responsibility for national park management and resource extraction regulation). During his brief tenure, James Watt proved to be a very controversial politician (see Bratton 1983). In order to facilitate the economic restructuring and rehabilitation desired by President Reagan, Watt fostered a pro-development attitude which sought to limit the state regulation of economic interventions within the natural environment (Coggins and Nagel 1990). This pro-development stance resulted in Watt supporting the use of federal property for the development of ranching and working to open up more areas of public land for resource exploitation. Watt also developed an increasingly antagonistic position towards environmentalists in the USA, famously describing them as a left-wing cult who where opposed to the basic values of American society.

Responses to Rio and the rise of the Wise-Use Movement

While Watt lost his job at the Department of the Interior in 1983, many of the principles of what the US environmental movement called Wattism continued in the USA. Perhaps the most influential development in these terms has been the emergence of the Wise-Use Movement (Harvey 1996: 383–385; Ehrlich and Ehrlich 1996). The Wise-Use Movement evolved in the American West during the late 1980s. Those in the Wise-Use Movement opposed the types of environmental regulations proposed by the EPA and other state institutions, on the basis that they were damaging local economies and associated job prospects and eroding the basic property rights of people. In this context the Wise-Use Movement has challenged the view that state-led forms of ecological modernization, or environmental regulation, are beneficial for economic development (a classic mantra of ecological modernization) and have proposed a very different vision of sustainability. The Wise-Use Movement consequently argue that private landowners always carry out the 'wisest-use' of any environment. They claim that landowners know and understand their land the best, and that it is in their own best interests to ensure its long-term sustainability. In this context, those involved in the Wise-Use Movement claim that excessive state intervention in land-use issues only serves to alienate people from the environment and further endanger its long-term sustainability (Harvey 1996: 384). In the context of this book, it is interesting to note the very spatial form which the Wise-Use Movement's politics has taken. While we have already talked about the importance of restructuring the spaces within which we live in order to create a more sustainable society, the Wise-Use Movement have used the notion of private space (or property), and its associated socio-economic values, as a way of questioning and resisting many of the ideals associated with state-sponsored sustainability.

At one level, the Wise-Use Movement has clear parallels with some of the principles of deregulated ecological modernization in the UK – at least in terms of realizing that you can never wholly force people to be sustainable. It differs, however, to the extent that it claims that little needs to be reformed within the existing patterns of economic practice and development. In certain extreme examples factions of the Wise-Use Movement go a stage further and argue that in order to preserve economic security and growth the environment should be exploited to a greater extent than is currently being experienced (perhaps in the form of logging and mining in national parks, or drilling in the Arctic) (Ehrlich and Ehrlich 1996). Of course as, Harvey (1996) points out, the liberalist logic of the Wise-Use Movement is partly flawed. It is flawed because despite the alluring and romantic appeal of the Wise-Use Movement's description of responsible small-scale landowners, large corporations and global companies – who are also major landholders – often have no real commitment or special knowledge of the properties they exploit. Indeed, it is in the best interests of such corporations to exploit certain areas to ecological breaking point and then simply move on. But, nevertheless, it is clear that the Wise-Use Movement (and its associated supporters) present a very different interpretation of sustainability, and how best to balance the needs of the economy and the environment, than those promoted within many state bureaucracies.

It is clear that Wattism and the values of the Pacific Legal Foundation and the Wise-Use Movement had an enduring influence on mainstream political attitudes towards sustainability in the USA during the 1990s. Perhaps the impact of such values was expressed most clearly at the Rio Earth Summit of 1992. While to many the Rio summit offered the hope of an epochal shift in international political values towards issues of environmental and social justice, the recalcitrant position adopted by the USA to many of the conference's policies was a major barrier to progress. Tim Luke (1999a) recognized two prevailing attitudes which characterized the US political stance at Rio. The first was an active concern with the economic consequences of the US government signing up to international agreements on climate change and biodiversity conservation. Many felt that to ratify such agreements would threaten the economic security and welfare of the USA and in particular damage the major oil and biotechnology sectors of the US domestic economy. Second, Luke (1999a) notices how in the build-up to and during the Rio summit itself, there was increasing suspicion in the USA over whether the nation should devote so much attention to issues of international justice, when there were pressing domestic matters to deal with back home. On these terms it is clear to see that despite the pervading processes of globalization, nation states and issues of national sovereignty are still crucial factors within any discussions of the sustainable society.

Following the Rio Earth Summit, there appeared to be new hope for the sustainable development agenda with the election of Bill Clinton and Al Gore. The Clinton–Gore partnership offered a sympathetic approach to issues of sustainability at a federal level, a point emphasized by the fact that while still a senator, Al Gore had written a book on the need for greater international consensus and action regarding global environmental threats – *Earth in the Balance: Forging a New Common Purpose* (Gore 1992). In keeping with their new-left (New Democrat) rhetoric, the Clinton administration argued for greater environmental reform at home (particularly in relation to the treatment and disposal of toxic substances) and a greater role for America in securing international environmental security (Luke 1999b). Essentially, the Clinton administration recognized that in the post-Cold War era, transnational security threats like those posed by environmental pollution and resource scarcities were themselves threats to the USA's own socio-environmental welfare. It appears that Clinton believed that the USA should use its resources and powers to act as a global environmental protection agency (Luke 1999b: 126). Despite the ostensible commitment to the principles of sustainable development at both home and abroad, however, in many ways the Clinton–Gore partnership remained a disappointment when it came to policies for sustainable development. On the domestic front Clinton approved an accelerated programme of drilling for oil and gas and failed to force car manufacturers to improve the fuel efficiency of their cars (Luke 1999b; Moore 2001: 210–211). Internationally the Clinton administration also failed to commit the USA to the Kyoto protocol on climate change, compromising the future of the international climate change programme (Klein 2001: 77).[7] It is important to emphasize that under Clinton a whole range of sustainable development policies were developed at both a state and federal level. Notwithstanding this, however, there was still an unwillingness within the Clinton administration to commit to certain international agreements regarding global sustainability.

Rowell argues that the political resistance to sustainability within certain parts of US society is part of a broader *green backlash* (Rowell 1996), or social resistance to the politics of environmental protection. Others have described it as a *brownlash* (Ehrlich and Ehrlich 1996), or the corporate–industrial response to the political development of environmentalism and the international social justice movement. However the 'lash' is understood, it is clear that in order to prosper within Western market economies, sustainability has to be compatible with prevailing economic systems. While (as we saw in Chapter 1) sustainability is all about making economic growth compatible with environmental protection and social justice, it appears that in Western states like the UK and USA, particular attention has been given to developing economically enthused visions of sustainability. Of course the danger with this emphasis on the economic viability of sustainability is that as soon as the political and economic credentials of sustainable development policies are questioned (as they have been by groups like the Pacific Legal Foundation and the Wise-Use Movement), there is a danger that sustainability itself can be dismissed as a model for future socio-environmental development.

Sustainable heterotopias: alternative Western experiments in sustainability

If the tone of this chapter seems overly pessimistic, perhaps this should come as little surprise. After all, sustainability does pose something of a 'threat' to Western liberal values of free enterprise, profit making and property ownership. Consequently when we look at the bastions of Western liberal values, namely government departments, parliaments, royal commissions and presidential offices, we should expect to find some suspicion and resistance to many of the values which sustainability implies. But if we look at some different spaces, more distant to the state apparatus in MEDCs, it is possible to discern a different set of circumstances emerging. Throughout his celebrated work, the famous French philosopher Michel Foucault emphasized the importance of analysing different, alternative and often taken-for-granted spaces when trying to unravel the prevailing logics of society, and attempting to uncover other ways in which a society could develop (see Hetherington 1997). Foucault referred to these alternative or 'other spaces' as *heterotopias*. The point is that if we scratch beneath the surface of the formal spaces of Western political society, we discover a whole range of alternative sustainable societies existing alongside ecological modernization, the Wise-Use Movement or Wattism.

Sustainable heterotopias vary in their form and function. For some they can be very private spaces, perhaps the domestic sphere, a private rural smallholding, or the practices by which people attempt to ensure that the products they consume and the resources they use during their everyday lives are utilized in a more sustainable way. Other heterotopias involve a marked separation from mainstream society and the creation of isolated worlds of sustainable living. Such spaces may include anarchist communes or biospherical experiments. Other sustainable heterotopias are devoted to promoting and educating people about other ways in which sustainability can be envisaged and delivered – good examples of these are perhaps the Centre for Alternative Technology in mid Wales (see Chapter 6) and the Los Angeles Eco-Village. While varying in their form, what many of these 'other spaces' of sustainability have in common is their rejection of Western values of capitalist economics, private property rights and profit making (see Routledge 1997). These heterotopias are often based on collective forms of ownership and participation, on making less, not more, money and on developing smaller-scaled modes of living which are more sensitive to the needs of nature and the surrounding environment. While often overlooked and ridiculed, these heterotopias are all around us and, unlike many of the state-based versions of sustainability we have discussed in this chapter, they do not assume that business can carry on as usual, but that society must radically alter its economic practices and socio-ecological values.

I will explore various examples of sustainable heterotopias in greater detail in Part II of this book. One particular alternative space of sustainability I want to describe here, however, is perhaps unusual. It is unusual because it is not a permanent space, but rather a fleeting occupation, or opening up of space for sustainability I witnessed several years ago. On 1 May 2000, a protest against Western capitalism involving over 4,000 people took place in the streets of London. The protesters were drawn from a range of political groups, including members of the Reclaim the Streets Movement, campaigning against the increasing privatization of pubic space, and people from the radical environmental and social justice movements (*The Guardian* 2000a). Following on from other anti-capitalist protests in Seattle and Washington, what appeared to unite the protesters was a common concern with the erosion of social and environmental values caused by the expansion of global capitalism. While the violent scenes with which the protests eventually ended caught the headlines, two things really interested me about the protest. First, I was interested in the way in which the protesters used the urban spaces of London as the basis for a kind of rallying call against capitalism. Second, I was intrigued with the significant material and ideological role which space appeared to play in the protests.

During the May Day protest, the urban environment was used as a basis for political action in a number of different ways. First, the protesters, many of whom were seasoned 'guerrilla gardeners' in their own right, sought to transform Parliament Square in central London into a more 'sustainable space'. The protesters stopped traffic from moving through the square and dug up parts of the green so that they could lay plants and seeds (*The Guardian* 2000b). In an act of classic *détournement*, the protesters also removed turf from the grassy areas of the square and placed it on the concrete roads which ran through the square. The use of environmental objects and practices in this way served to illustrate the problems associated with the kinds of urban environments produced under capitalism. As one protester emphasized at the time:

> It's not just green; it's also symbolic of taking back the land under people's feet – land they have to buy or rent – and breaking up the alienating cycle of producers and consumers. It represents the idea that people have the power and right to decide how they live. We do not have to live in this ugly and constrained way under capitalism.
>
> (Jack Tann, Reclaim the Streets, *The Guardian* 2000b)

Essentially the kinds of environmental actions of the May Day protesters represent an alternative, albeit far from coherent, vision of a sustainable society – one which brings into question the basic values of many MEDCs.

Another significant aspect of the May Day protests was the way in which the protesters used space as a critical arena through which to contest the socio-environmental consequences of capital. As part of their protests, the campaigners symbolically occupied a very important space of representation to the British nation – Parliament Square. The use of Parliament Square was a deliberate act of spatial resistance on behalf of the protesters

(Cresswell 1996). Located next to the Houses of Parliament (the seat of British government and the symbolic heart of the UK state) and Westminster Abbey (a prominent symbol of the Anglican Church in Britain), Parliament Square is a significant space in the British national imaginary. Containing the only statue of Sir Winston Churchill in London, the square is a celebration of British statehood, a space upon which the state projects its own image. In occupying, cultivating and 'disrupting' this space, the protesters were directly challenging the social, economic and political values of the state. The physical and symbolic importance of space within the protests was echoed in the sentiments of one columnist at the time:

> What they [the protesters] have is the leverage to open space for instrumental action against polluters and profiteers, and sometimes just to open space so that we can remember what open space looks like.
>
> (*The Guardian* 2000a)

What the London protests appear to reflect, albeit it in a very ephemeral way, is that space is a key arena through which resistances and challenges can be made to the dominant forms of social and economic development we see around us today.

Summary

In this chapter we have explored the modes of sustainable development which have been emerging in certain MEDCs over the last thirty-five years. It is clear that in the Western democracies we have considered, there has been a concerted desire to forge a brand of sustainability which is broadly compatible with the social and economic values of neo-liberalism and free-market economics. In the UK, the principles of ecological modernization have provided a framework within which to convince businesses that policies for sustainable development actually benefit, not inhibit, economic growth. In the case of the USA, we have seen how despite attempts to develop modes of sustainability which are acceptable to the business community, sustainability can often challenge the values of property-based economic liberalism which are woven into the fabric of Western society. Where this has happened in the USA, there has been something of a green backlash (or brownlash) which is now making the broader instigation of the principles of sustainability into American society increasingly difficult.

I do not wish to argue that what is happening in the UK and USA are necessarily typical MEDC interpretations of sustainability – in Japan, for example, very different versions of a sustainable society, based upon technological innovation and excellence, are emerging. However, it is clear that the tensions which will always surround attempts to make sustainability acceptable to the hyper-industrialized economic sectors of Western nations represent a key challenge for advocates of sustainability as we move into the twenty-first century. I argue that the alternative spaces (or heterotopias) of sustainability in MEDCs are important in this context. They are important because they question the assumption that sustainability should automatically be something which can be easily synthesized into existing economic practices. Perhaps sustainability should be something which challenges us to rethink our economic values and practices, rather than allowing existing economic structures to shape our social and environmental policies.

Suggested reading

For a detailed description of ecological modernization I would recommend Dryzek, J. and Schlosberg, D. (2001) *Debating the Earth: The Environmental Politics Reader,* (part 3, especially section VII) and Hajer, M. (1997) *The Politics of Environmental Discourse: Ecological Modernization and the Policy Process.*

For a concise but excellent introductory critique of the Wise-Use Movement, see Harvey, D. (1996) *Justice, Nature and the Geography of Difference*: 383–385.

Finally for a more general overview of contemporary political resistances to environmentalism in both the West and less economically developed countries see Rowell, A. (1996) *Green Backlash: Global Subversion of the Environment Movement.*

Suggested websites

Here are website addresses for the key institutes I have mentioned in this chapter:

Carbon Trust: http://www.thecarbontrust.co.uk/carbontrust/

Department of the Environment, Food and Rural Affairs: http://www.defra.gov.uk/

Environmental Protection Agency: http://www.epa.gov/

The Pacific Legal Foundation: http://www.pacificlegal.org/

For those of you interested in exploring some sustainable heterotopias, I can also recommend visiting the following sites:

Biosphere 2: http://www.bio2.com/index.html

Centre for Alternative Technology: http://www.cat.org.uk

Los Angeles Eco-Village: http://www.ic.org/laev/

3 SUSTAINABLE DEVELOPMENT IN THE POST-SOCIALIST WORLD

Introduction

I will always remember the first time I visited Katowice. I was driving from Wrocław in south-east Poland in a large white Land Rover which I had been allowed to use by the Gordon Foundation. I was in Poland studying environmental movements which had emerged in the socialist and post-socialist era in the country. I noted that in the *Rough Guide to Poland* the city of Katowice had been rather unfairly characterized as having little to detain the tourist. I was, and still am, fascinated by Katowice, however, because of the ways it embodies the social, economic and environmental struggles which have characterized Poland's wider shift from socialism to liberal democracy and capitalism. The road between Wrocław and Katowice provided a fascinating insight into the recent history of the country. For much of the journey I was glad to have the use of a Land Rover, and its heavy-duty suspension, as we travelled across the eroded concrete road surfaces which had been laid and poorly maintained during the communist era. Occasionally poor road surfaces would make way for the smooth tarmac of Poland's new European Union-sponsored roads, an indication of its post-socialist transition to EU member state. Alongside the road we passed endless high-rise tower blocks, used under the socialist government to house the workers of the state-run urban industries, interspersed by new out-of-town shopping centres, replete with McDonald's fast-food joints and IKEA stores. As we approached Katowice we saw increasing evidence of the environmental legacies of state socialism in the region. The scorched landscapes associated with mining and the billowing smoke plumes of heavy industrial works served to remind us of the huge environmental costs associated with the economic plans which had been enforced upon Katowice's regional economy. Despite these portentous signs of environmental decline, however, I was actually travelling to Katowice to study its recent innovative attempts to create a more socially and ecologically sustainable city through a new system of integrated municipal management.

What Katowice, and my journey to it, revealed were the complexities and contradictions of the post-socialist world. To speak of 'the post-socialist world' – as I do throughout this chapter – is convenient, but also misleading. It is misleading for two main reasons. First, because it suggests that countries which have recently rejected state socialism have now left socialism and its associated social and economic systems behind. As my journey through Poland revealed, however, post-socialist societies still exist in the social and political shadow of socialism and still carry its socio-environmental and political legacies. Second, to speak of a post-socialist world is misguided because it tends to create a picture of a single, post-socialist society, on a single developmental transition away from socialism and towards capitalism. In reality, as many writings about the geographies of post-socialist society have revealed, different states and regions have adopted very different socio-economic modes of post-socialist existence, and have developed very different transition strategies and trajectories from socialism (see Bradshaw and Stenning 2004: chapter 1).

Transition is a word which appears regularly in writings on post-socialist society (Pavlínek and Pickles 2000: chapter 1). With its etymological association with notions of change and transformation in time, *transition* has been used to understand a range of different components of post-socialist society: *transition economies*; *transition societies*; and even *environmental transitions*. While the notion of transition usefully conveys the huge upheavals which have been experienced in the post-socialist world over the last twenty years, the idea of transition can be problematic. At one level the idea of transition is troublesome because, as I have already mentioned, it can be interpreted as referring to a singular process which all post-socialist countries are experiencing. According to David Stark (1992), however, this simplistic idea of transition does not do justice to the plurality of different transitions which are occurring within different post-socialist states and regions. Transition it would appear is a plural not a singular process, which varies in both its form and speed between different states and even within nations themselves (ibid.). The notion of transition is also problematic because while capturing the fluidity and flux of post-socialism, it fails to encapsulate the inertia of socialism and the associated political, economic and environmental legacies of socialism that clearly persist. One thing which writers on post-socialist transitions appear to agree on, however, is the systematic nature of transition in the post-socialist world. Post-socialist theorists have deployed the idea of *systematic transformation* to capture the multidimensional nature of change after socialism (see Bradshaw and Stenning 2004: chapter 1). In this context, it is argued that post-socialist transition cannot be reduced simply to economic change, but must be understood as an overhaul of the entire socio-political and cultural infrastructure of former socialist states. Significantly, as we will see, the restructuring of their social, economic and environmental systems has given post-socialist communities the opportunities to incorporate the principles of sustainability into their emerging development plans. It is the relationship between post-socialist transitions and sustainability which provides the focus for this chapter. But, as we will see, the different uses and interpretations of sustainability within post-socialist states is as varied as the transitional approaches they are encountering.

Analysis in this chapter considers the relationships between socialism, post-socialism and sustainability specifically within Central and Eastern Europe and Russia. This chapter begins by considering the kinds of socio-environmental relations and economic system which characterized the socialist era. In the context of these legacies, the following section considers how and why the principles of sustainability have been incorporated into post-socialist development plans. The remainder of this chapter then focuses on two case studies. The first considers socio-environmental relations in socialist and post-socialist Katowice (Poland). The second explores emerging sustainable development policies in Russia. Ultimately this chapter seeks to illustrate the different geographical forces which are shaping sustainable development in the post-socialist world and the opportunities and constraints facing sustainable development in transition societies.

Socio-environmental legacies of socialism

Understanding socialism

> 'Did not Lenin himself admit that communism was socialism plus electricity?'
>
> (Lewycka 2005: 87)

As with most phrases prefixed with the word *post*, it is important to consider what the socialism within *post-socialism* actually is before we even begin to consider what came after it. As a phrase socialism has become synonymous with two related terms – those of Marxism and communism. It is useful to start our analysis of socialism by considering these. While notions of socialism pre-date the extensive and influential works of Karl Marx, it is clear that Marx has provided the most comprehensive set of analyses and accounts of socialism. As an intellectual tradition Marxism was and is primarily concerned with the unfolding socio-economic relations associated with capitalism. In his famous three-volume analysis of the inner workings of the capitalist system, *Das Kapital*, Marx exposed the social and economic flaws of capitalism. Perhaps the most crucial dimension of Marx's analysis of capitalism was his account of the newly emerging class of proletariat. The proletariat, or working class, were workers who underpinned the industrial revolution. According to Marx, however, while producing the wealth associated with the industrial revolution, the proletariat experienced few of its benefits, suffering from extremely low pay, poor housing and poverty. Meanwhile, the owners of industry gained new wealth and opulence in their own lifestyles. Marx's analysis of the proletariat exposed the great social injustices associated with capitalism, but it was also used by Marx to reveal the path to a post-capitalist future – namely socialism. In their *Communist Manifesto* (1848), Marx and Engels described how the expanding, but ever disenfranchised, working class must seize political power in order to create a fairer, more just society.

It is the idea of a working-class government – or *dictatorship of the proletariat* – which was the leitmotif of socialism. As a political system, socialism was to be based upon a centrally planned and administered, single-party state, centred on the socialist party. Such a political system was seen as being vital to ensuring a more equal distribution of goods and welfare throughout society than had been achieved under capitalism. In this context the socialist state sought to follow the simple but famous dictum: *from each according to their abilities, to each according to their needs*. The idea of a single-party state was crucial to socialist values, because it was thought that any more than one political party (or indeed social movement or free trade union) would inevitably mean that one party, group or union would favour the interests of capital more than the others. Only the workers party could avoid this. Consequently, even where other political organizations were allowed to exist in socialist societies (like nature clubs and youth groups), these groups were always carefully monitored and regulated by the state. Beyond these political structures, socialist society was also based upon a very different set of economic values than are common under capitalism. The distinctiveness of socialist economic systems was largely determined by the relationship between the state and economy. The economies associated with socialist states have been described by Gregory and Stuart (1998: 99) as, 'administered command economies' (also see Bradshaw and Stenning 2004: chapter 1). The idea of a carefully administered economy eloquently conveys the key difference between socialist and capitalist economic systems. Consequently, while capitalism is characterized by the operation of the free market and only limited forms of state intervention, under socialism the state dominates the economy. The state control of socialist economies was really achieved through two key processes. First, through the nationalization of key industries and property rights, socialist states were able to take direct control of key sectors of the economy (see Whitehead 2005: 277). Second, through the use of economic plans, targets and goals, socialist states sought to control the growth and outputs of key industries. As Bradshaw and Stenning (2004: 7) note, despite the state control of key industries, large amounts of economic practice still existed outside the state monopoly sector. Despite this, however, through the use of nationalized industries and economic planning, socialist states endeavoured to replace the forces of the free market with an economic system which was designed to meet the political needs and demands of the state (ibid.: 6). Interestingly, in the work of Marx (and Engels) it is clear that the socialism of planned state economies, and the dictatorship of the proletariat, was only to be a staging post or point of passage on the path of transformation between capitalism and communism. In this context communism was seen as the ultimate classless society – a society of the future, within which the redistributions enacted by socialist states would enable the dissolving of the state and the rise of communal living and working.

If the work of Marx and Engels contained the ideals upon which socialist society was to be based, the rise of socialist states was something which Marx and Engels would not themselves see (although for an interesting discussion of the local rise of socialism in the nineteenth century see Harvey 2003b). It was not until Russia's October revolution in 1917 that a large-scale socialist state became a political reality. Following the Bolshevik revolution of 1917, socialism gradually spread from Russia, and later Soviet states, to include large parts of Central and Eastern Europe following the end of the Second World War. Subsequent revolutions during the second half of the twentieth century saw the geographical spread of socialism into Asia (China, North Korea, Laos and Vietnam for example), Africa (Burundi, Guinea and Mali among others) and the Caribbean (Cuba). Despite the rapid geographical spread of socialism during the twentieth century, however, many still question whether the true ideal of a socialist society has ever been implemented as a form of government.

Actually existing socialism and socio-ecological crises

The difference between idealist visions of a communist world and reality is perhaps best captured in the phrase *actually existing socialism*. This phrase (or *real socialism* as it is referred to in some post-socialist states) denotes the types of political, economic, social and environmental systems which emerged in the socialist world following socialist revolution and the ascent of socialist parties. In Russia and Central and Eastern Europe, for example, the principles of Marx were interpreted and applied in practical political situations by Lenin and then Stalin, which gave rise to a distinctively Soviet style of socialist society. Rather ironically, many writers have argued that this style of socialism was so far removed from the ideal vision of communism that it was actually more akin to a refined capitalist state system. In his famous analysis of communist state systems in Russia, Tony Cliff (1974) coined the term *state monopoly capitalism* to describe the actually existing socialisms he encountered. According to Cliff, under Soviet-style socialism, the class interests of the bourgeoisie (the owners of capitalist industry) were replaced by the monopolistic interest of a new political class of state managers. With a desire to ensure the economic success of socialism – particularly relative to capitalism – the new class goal associated with socialism became the expansion of national income. While this class goal was different to the expansion of private class incomes under capitalism, it resulted in a system of economic expansion and the consolidation of elite class interests which was remarkably similar to free-market capitalism (see Whitehead 2005: 277). Cliff and others have argued that the main difference between capitalism and state monopoly capitalism was that while capitalism was based upon private property relations, socialism was predicated upon the politically unaccountable, public exploitation of economic resources (Kuron and Modzelewski 1982).

In the context of these links between capitalism and actually existing socialism, you may expect the socio-economic and environmental consequences of both systems to be relatively similar. Many scholars now argue that the consequences of actually existing socialism were, however, much worse than those of free-market economies. At a social level numerous harrowing accounts have emerged describing the difficult nature of social life in the socialist world (see Amis 2002; Solzhenitsyn 2002). Traditional ways of life were shattered under actually existing socialism as large-scale urban industrial expansion was implemented as a means to maximize national income. For those moving to industrial cities there were rewards, as industrial workers were glorified and rewarded for their efforts for supporting the national economy (see Bialasiewicz 2002). Stenning (2005: 116–117), for example, is quick to point out the social benefits and emerging systems of *local patriotism* which characterized urban industrial expansion under socialism. Nevertheless it is clear that these *hero workers* had to endure heavily overcrowded and polluted living conditions in newly expanding socialist cities – conditions which betrayed the promises of communist ideology (Bradshaw and Stenning 2004: 6) (see Figure 3.1). Meanwhile those left in rural areas were to endure the socio-cultural turmoil associated with the collectivization of agriculture and in certain extreme cases the terror famines which were to become such a notorious component of Stalin's European regime.

The social problems associated with the socialist world would, however, perhaps have been more politically manageable if related economic systems had operated effectively. Successive analyses of Soviet-style economics have revealed the significant flaws in the structures and practices of socialist states. One of the key problems associated with systems

Figure 3.1 Mass housing complex, Katowice, Poland

of state monopoly capitalism was that while national economic plans put a premium on high and ever-increasing economic output, little attention was given to the costs which were associated with accelerating industrial production (see Bradshaw and Stenning 2004: 7). Without sufficient attention given to the need to improve economic efficiency (both in relation to labour practices and industrial infrastructure), it became difficult to sustain economic growth (Khanin 1992). By the early 1980s, it was clear that an economic crisis was facing many socialist states. This would of course lead to the iconic pictures of bread queues which became the precursor to market-style economic reform in socialist states in Europe and the former Soviet Union. In many ways the emerging economic crisis of socialist states in Central and Eastern Europe and Russia was interconnected with the environmental problems of socialism.

The environmental crisis of socialist societies in Central and Eastern Europe and the former Soviet Union has been well documented by a range of geographers, sociologists and environmental scientists (see for example Oldfield 1999; Pavlínek and Pickles 2000; Peterson 1993). It appears that the intensification of economic development and hyper-industrialization led to widespread environmental degradation and pollution in many socialist states. Successive studies have shown that European socialist states were among the world's worst atmospheric polluters (see Figure 3.2) (Pavlínek and Pickles 2000). Allied to this, many former socialist states suffered from severe water pollution and land degradation, both of

Figure 3.2 Air pollution in Upper Silesia, Poland

which were associated with the rapid spread of mining (see Box 3.1). The environmental legacies of socialism were clearly the product of the political and economic systems which became characteristic of actually existing socialism. Economically, environmental pollution was the product of both the pressures to continually increase national economic production and the lack of attention given to the need to increase industrial efficiency through new forms of technological investment in the production process. Politically it is also clear that the monopolistic nature of socialist party state rule contributed to the environmental crisis of *real socialism*. Because of the dominance of the socialist party within systems of socialism, it was very difficult for green or environmental movements to emerge in the socialist world. Where environmental groups did exist, they were either politically conformist nature clubs, or underground protest movements with relatively little formal political influence (see Hicks 1996). In the relative absence of large environmental protest groups, very little pressure could be brought to bear on socialist governments to reform their environmental policies. This situation was also compounded by the lack of environmental pollution data which was made publicly available by socialist states. In certain states it appears that the production of pollution data was not a high priority (see Pavlínek and Pickles 2000). Even when socialist states produced good environmental information, it was often keenly guarded by government officials (see Jehlicka and Cowell 2003). With limited environmental protests and environmental pollution data, certain commentators have argued that unfettered environmental pollution was able to endure much longer in the socialist world than it was in the West.

Box 3.1 The environmental crisis of actually existing socialism: case study of the Black Triangle

A classic example of the types of environmental problems generated by actually existing socialism is provided by the case of the Black Triangle. The Black Triangle is an unofficial region which has experienced some of the worst cases of pollution in Europe. There are two reasons why this region is referred to as the Black Triangle. The reference to 'triangle' is because the region in question cuts across three countries – south-west Poland, northern Czechoslovakia (now the Czech Republic), and what was East Germany. It is known as the 'black' triangle because of its reputation for producing exceptionally high levels of black smoke. According to Nowicki (1993: 22 and 106), towards the end of the 1980s the Black Triangle region was responsible for producing three million tonnes of sulphur dioxide every year – that represented a staggering 20 per cent of all of Europe's sulphur dioxide output (Pavlínek and Pickles 2000: 48). The reason for these extraordinarily high pollution levels was that the Black Triangle was home to a valuable set of resources (particularly brown coal). In this context, the area was prioritized within the economic plans of the East German, Polish and Czechoslovakian states. Because the Black Triangle was seen to be vital to the economic success of different socialist states, it quickly became home to one of the world's largest concentrations of brown coal energy plants (Pavlínek and Pickles 2000: 47–48). The region was also home to a large quantity of opencast mines, which resulted in a heavily scared and despoiled landscape. One of the only responses to try and address the problems of environmental pollution in the region was to build higher smoke chimneys so that air pollution could be more effectively dispersed away from the area (see Figure 3.3).

Key reading: for a very detailed account of the environmental problems of the Black Triangle see Pavlínek, P. and Pickles, J. (2000) *Environmental Transitions*: chapter 3.

Figure 3.3 A smoke dispersal chimney, Upper Silesia, Poland

The (re-)emergence of sustainability in post-socialist society

'Was it better under communism?' I ask.

'Of course better. Was good life. You no understand what type of people is running country now.'

(Lewycka 2005: 112)

Understanding the social, economic and environmental relations of socialist state systems is important if we are going to effectively interpret and analyse the post-socialist world and the emergence of sustainability therein. An appreciation of the socialist world provides an important context within which we can begin to understand how society has been changing in the post-socialist era. In recognizing this, however, care must be taken in how we understand the relationship between socialist and post-socialist societies. As we have already discussed at the beginning of this chapter, the relationship between socialism and post-socialism is routinely understood through the notion of *transition*. Recent theories of transition, as we will see, do, however, challenge many of the assumptions which lie behind analyses of post-socialism.

Thinking through post-socialist transitions

Within work on post-socialist societies, there is something which is called an *orthodox approach* to transition (Bradshaw and Stenning 2004: 12). This orthodox perspective, while (sometimes) acknowledging the diversity and complexity of the post-socialist transitions which are occurring in different post-socialist states, argues that there is a basic underlying logic to transition. According to the orthodox perspective then, post-socialist transitions involve the abandonment of one mode of social existence – the socialist planned economy – which is replaced completely by a new political economic system – namely free-market capitalism (ibid.: 8). While adopted by certain scholars, this orthodox position is also characteristic of the outlook of the World Bank and other key international institutions involved in shaping the future of the post-socialist world. According to Bradshaw and Stenning (2004), there are at least three major problems associated with the assumptions underlying the orthodox model of post-socialist transition. First, they claim that the orthodox position suggests that there is a thing called socialism which is to be left behind, when in fact there are a range of different socialist traditions which existed in the socialist world. Second, they argue that this paradigm of transition over-simplifies the goal of free-market capitalism that post-socialist societies are supposed to be moving towards (ibid.: 13). In opposition to this claim, many now argue that there is no single dominant, or monolithic, global capitalist system, but rather a complex web of more or less capitalist economies (see Gibson-Graham 1996). Thirdly, and finally, Bradshaw and Stenning claim that the orthodox approach tends to see socialism as being completely erased by free-market capitalism. Research clearly shows that in many post-socialist states there are many socio-cultural, political and economic traditions and practices which have endured since the socialist era. Consequently, while the received wisdom states that the collapse of the Berlin wall and the tumultuous political events of 1989 led to what Fukuyama (1993) has described as the *end of history* (and in particular the end of the Cold War struggle between capitalism and socialism), this accepted narrative is misleading. What instead has been emerging in post-socialist countries are a series of proto-capitalist systems, which are heavily infused with certain socialist traditions, institutions and practices. It is these hybrid political and economic forms which anyone writing on or working within post-socialist communities must recognize and adapt to.

Relationships between post-socialist transitions and sustainability

Questions of sustainability appear to have been an important factor in influencing and guiding transition societies in Central and Eastern Europe and

Russia. At one level, for example, it is clear that issues of sustainability were crucial in the emerging crises of socialist societies. Whether it be in relation to the difficult social conditions experienced in socialist states, failing economies, or the destruction of environmental assets, one of the main reasons why political protest groups started to question the validity of socialist state systems was on the fundamental basis of how sustainable they were (see Pavlínek and Pickles 2000: part II). In addition to providing some of the impetus for post-socialist transformations in the first instance, sustainable development and different forms of sustainability have played an increasingly important role in the emerging political, economic, social and environmental systems associated with post-socialism. It appears that the socio-political upheavals associated with transition societies have provided a unique opportunity for the incorporation of the principles of sustainability into the workings of post-socialism. Jonathan Oldfield (2001: 105), writing about sustainable development in Russia, observes how some people believe that sustainable development has actually provided an alternative social and moral model to socialism in the post-socialist world. Whether this is the case or not, it is clear that the principles of sustainability have been enshrined within a range of different laws, policies and planning frameworks, which have collectively laid the foundations for the form which different transitional societies have taken (see Figure 3.4).

Significantly, what geographical work on the emergence of sustainability in post-socialist societies has revealed, is that the agreed international principles of sustainable development have not simply been

Figure 3.4 A green energy project in Wrocław, Poland

translated into post-socialist contexts (see Baker 2000; Baker and Jehlicka 1998; Baker *et al.* 1997; Jehlicka 2001; Jehlicka and Kostelecky 2003; Jehlicka and Tickle 2004; Jehlicka *et al.* 2005). Instead, it is apparent that sustainability has encountered pre-existing communist and pre-communist socio-ecological and cultural–environmental traditions, and has combined with these to form new patterns and models of sustainability (see case studies below) (see also Oldfield and Shaw 2002). Despite the new opportunities offered for the incorporation of the principles of sustainability in to post-socialist societies, it is clear that other pressures associated with transition have inhibited the development of effective policies for sustainable development. Prime among these barriers has been the problems and pressures associated with economic transformation. Following the collapse of state socialism in Europe and the Soviet Union, something referred to as the *transitional recession* swept throughout the post-socialist world (Bradshaw and Stenning 2004: 20–21). The transitional recession was the term used to describe the rapid slow down of economic activities in post-socialist states following the collapse of the planned economic system. The reasons for the transitional recession are complex, but largely relate to the collapse of the traditional markets and trading zones of the socialist era and the uncompetitive nature of post-socialist economies compared to their Western neighbours. Despite the causes of economic recession, it is clear that throughout much of the post-socialist world the urgent need for economic reform has overshadowed the drive towards sustainability, and clearly inhibited the emergence of sustainable development. The remainder of this chapter will explore the complexities surrounding the relationship between post-socialist society and sustainability through two geographical case studies: the first considers the post-socialist city of Katowice in Poland and the second explores the emergence of sustainable development in Russia.

Exploring sustainability in the post-socialist world I: Sustainable development in the post-socialist city

Socialist Katowice and the unsustainable city

The first example of the relationship between sustainability and post-socialist transition I am going to consider is Katowice – the Polish city I described travelling to at the beginning of this chapter. Cities, and their transformation under post-socialism, have been the subject of much academic debate and conjecture (see Andrusz *et al.* 1996). As loci of socio-economic activity and intense meeting points of society and nature, cities have provided important insights in to the nature of post-socialism.

Katowice (or the Katowice region as it is sometimes referred to) is Poland's largest urban–industrial agglomeration, constituting fifteen adjoining towns and a population in excess of two million people. Katowice is located in southern Poland, in the region of Silesia (Figure 3.5). The relationship between Katowice and its surrounding regional hinterland lies at the heart of the story of the city's transition from socialism and its subsequent embracing of the principles of sustainability. Silesia is home to one of the most abundant concentrations of industrial resources in the whole of Europe (particularly in relation to coal and metal ores). It was in this context that Katowice quickly became the key centre through which the industrial assets of Silesia were transformed into the economic products which would drive the Polish economy.

Because of its wealth of mineral resources, Silesia was the object of concerted geopolitical struggles during the twentieth century. Towards the end of the Second World War, however, the Red Army occupied the region

Figure 3.5 Map of Katowice agglomeration

and control over the region was eventually passed to the newly formed Polish socialist state in 1946 (see Pounds 1958: 119–206). Once in the possession of the socialist state, Silesia became an important part of the nationally planned economy, with considerable emphasis being placed on the region as a source of national wealth. Successive regional plans for Silesia gradually saw the concentration of mines, energy plants and metal works in Katowice. As the economic infrastructure of Katowice expanded, so too did its population, as more and more people moved from the countryside and other regions in Poland to take up relatively well-paid jobs in the sprawling metropolis. In her compelling analysis of Upper Silesia, Bialasiewicz (2002: 117) describes the emerging popularity of the region in the post-war years, as promises of better housing, food and even holiday homes made the area seem like a 'worker's Eldorado'. It was in relation to such processes that Katowice rapidly became associated with the ideal dream of socialist progress, as the destructive spaces of urban capitalism were gradually replaced by the socially and ecologically harmonious spaces of the planned economy (see Whitehead 2005).

The ideal dream of socialist urban development in Katowice was, however, rapidly replaced by a much more disturbing *actually existing socialist city*. Because of the planned hyper-industrialization of Katowice, the urban agglomeration quickly became a key centre of Polish national economic development. Indeed by the end of socialism in Poland Katowice had 4,400 operational industrial plants (15 per cent of all Poland's plants) and was responsible for the production of 56.6 per cent of the nation's raw steel, 97.6 per cent of the country's hard coal and 100 per cent of Poland's lead and zinc (Katowice Voivodship 1991 in Nawrocki and Szczepanski 1995). When all these economic statistics were analysed together, it was estimated that the Katowice agglomeration was at times contributing 17.8 per cent of Poland's gross domestic product (Nawrocki and Szczepanski 1995). While such economic activity was good for the socio-economic needs of the socialist state, it was to have severe consequences for the socio-environmental conditions associated with life in Katowice itself. Environmentally, the concentration of industrial activity in and around Katowice was devastating. The atmospheric pollutants produced by Katowice's industrial plants meant that in 1994 the city was producing levels of black smoke pollution which were six times higher than those permissible under European Union legislation (Nawrocki and Szczepanski 1995; OECD 1994). The extensive patterns of mining in Silesia also led to the despoliation of approximately 20,000 ha of land in the late 1980s alone (Kabala 1991; Nawrocki and Szczepanski 1995; Pavlínek and Pickles 2000). High rates of industrial development, combined with an increasing resident population in Katowice, also led to an emerging waste management crisis in the agglomeration. The production of large amounts of industrial and domestic waste in the urban region, combined with the lack of a city-wide governmental body, made integrated waste treatment difficult to achieve and resulted in a water pollution crisis. The environmental problems associated with Katowice were compounded by related social problems in the metropolis. In addition to the often-poor quality working and living conditions experienced by residents, the environmental problems of the city gradually resulted in a deteriorating health record. For example the Provincial Hygiene and Epidemiology Station for the Katowice Voivodship – the body responsible for monitoring health rates and diseases in the city – estimated that in 1991 Katowice was experiencing incidence of cancers, heart disease and chromosome damage among young people that were far in excess of the national average (Nawrocki and Szczepanski 1995: 29–30) (for a broad review of the socio-environmental problems of Katowice see Whitehead 2005).

In essence what this collection of social and environmental statistics reveal is the betrayal of the socialist city ideal and the rise of the actually existing socialist city under modes of state monopoly capitalism. For a long time the inducements offered to key industrial workers in Katowice, and the lack of political opportunity to express concerns over socio-environmental welfare in the city, meant that the unsustainable pattern of urban development was able to endure. It was in this context that Katowice and Upper Silesia became a prominent pillar of the polluted Black Triangle region (see Box 3.1). Gradually, however, the internal politics of the city started to change. Interestingly, in the context of this book, the problems and emerging solutions to socio-environmental crises facing Katowice cannot be understood as products of either the city or the wider communist system alone. Western markets have long been consumers of cheaply produced goods from countries such as Poland, indirectly supporting the socio-environmental injustices which lay behind the production of such commodities. More recently, however, and through the auspices of the United Nations and European Union, Western governments have supported political and economic reform in cities like Katowice and promoted more sustainable patterns of development (see Baker 2000; Jehlicka and Tickle 2004). It appears that when it comes to discussions of sustainability it is important to remain conscious of the complexity and relative openness of contemporary geographical relations.

Transition and sustainability in the post-socialist city: achieving solidarity with nature

It is tempting to relate the collapse of socialism in Poland with the types of social, economic and environmental conditions experienced daily by those living in Katowice. In reality, of course, the collapse of socialism was the product of both macro-economic and large-scale political change, as well as the local everyday struggles experienced by those living in the socialist world. Nevertheless, it is clear that the socio-economic and environmental discontents of places like Katowice were one among a number of different reasons why Polish people started to question the socialist state system in the country. Poland was actually home to one of the most prolonged and successful struggles against Soviet domination in Central and Eastern Europe. This struggle focused upon Solidarity – a free trade union which was created to defend the rights of Polish workers (see Box 3.2). As we have already discussed, socialist ideals were fundamentally

Box 3.2 Solidarity and political change in Poland

Solidarity (*Solidarnosc*) emerged gradually in Poland as a loose association of workers' movements concerned with the treatment of labourers. Anxieties over workers' rights in Poland became a major issue in 1976 when striking workers were held captive by the state as punishment for their actions. In the context of such suppression, many activists and intellectuals felt that a free trade union was required to support the needs and rights of Polish workers and to avoid a repeat of what happened in 1976. Many consequently claim that the gradual emergence of the Solidarity movement began with the creation of the *Workers' Defense Committee* in 1976 (see MacShane 1981; Urban-Klaehn 2005). Solidarity's defining political moment and point of origin came in August 1980 at the Gdansk shipyard in northern Poland. Following a series of strikes over the summer of 1980, Lech Walesa was fired from the shipyard because of his role in workers' disputes. A strike committee was formed at the shipyard to demand Walesa's re-instatement, but because the strikers' committee were unable to reach an agreement with the shipyard, strikes started to break out all over Poland (including miners' strikes in Silesia and around Katowice) (Urban-Klaehn 2005). Eventually state negotiators were called in to resolve the disputes. Lech Walesa famously signed an agreement on 31 August 1980, which ensured better pay for workers and also secured the release of political prisoners (ibid.). The agreement also allowed for the formation of Poland's first free and independent trade union – Solidarity.

Key reading: Urban-Klaehn, J. (2005) *A Brief History of the Solidarity Workers' Union*. Available at: http://www.bellaonline.com/articles/art 34963.asp; Kuron, J. and Modzelewski, K. (1982) *Open Letter to the Party*; MacShane, D. (1981) *Solidarity: Poland's Independent Trade Union*.

opposed to forms of political representation which were outside of the socialist party. The relative success of Solidarity in securing better pay and working conditions throughout Poland provided a political context within which a range of new political movements were able to flourish. This era of emerging political freedom and opportunities in Poland is often referred to as the *Solidarity period*, and ran from the late 1970s throughout the 1980s. While the Solidarity period was still marked by strict socialist political rule, the relative political freedoms of this time meant that in many ways Poland was far in advance of other socialist states in beginning to imagine a post-socialist country.

The effects of the Solidarity period can be clearly observed in Katowice. Despite the strong emphasis which had been placed on hero workers and working privileges in Katowice, during the 1980s it became an active centre for the Solidarity movement. Those associated with Solidarity in Katowice started to question the working and living conditions which the industrial employers and the local government were providing (Bialasiewicz 2002). In the shadow of the Solidarity movement, other groups gradually started to emerge in Katowice. Prominent among these groups was the Silesian Ecology Movement. According to Tomeczek (1993), and other Polish scholars writing at the time, the Silesian Ecology Movement formally started in 1987 at a seminar held in Gliwice. The movement was a collection of students and civil rights groups who were concerned with the treatment of nature in and around the Katowice agglomeration (see Whitehead 2005: 287). What the movement essentially called for was research into the nature and extent of environmental damage in Silesia and the development of a clear strategy for protecting and enhancing the environment in the future (Tomeczek 1993: 163). The Silesian Ecology Movement challenged the socialist exploitation of nature through the regular publication of *Dead Nature* (an ecological magazine), marches, public speeches and mass demonstrations. While I do not want to claim that either Solidarity or the Silesian Ecology Movement was directly responsible for the transition from socialist to post-socialist society in Poland, I do argue that in Katowice at least, both movements questioned the prevailing socio-ecological orthodoxies of socialism and established some of the parameters upon which a post-socialist city should be founded (for more on the role of environmental movements in political transition in Central and Eastern Europe see Jehlicka 2001).

Following the collapse of socialism in Poland there has been a rapid uptake of the principles of sustainable development in the country. In environmental terms, Poland's embracing of sustainability was vital to its accession to full membership of the European Union and its continued transition away from socialism and Russian influence (see Baker 2000: 161). In Katowice it is possible to discern the local adoption of sustainability as a model for post-socialist city development. Given the severe social and environmental problems of the city under socialism, it was felt by local political leaders and urban residents that future development in the city should take much greater account of the everyday needs of people and understand how these needs are intimately tied to a more just and respectful use and treatment of the environment. The most direct expression of sustainability in the Katowice agglomeration is the United Nations-sponsored Sustainable Cities Programme there. Following the signing of an inter-municipality agreement between the different towns of the Katowice agglomeration, the Sustainable Katowice Agglomeration Project (hereafter SKAP) began. Through extensive public consultation the SKAP sought to lay the foundations for a more socially just and environmentally sustainable future for the city. At the centre of this project was a desire to unify an urban planning system which had become increasingly fragmented under socialist rule. Under socialism, for example, it wasn't possible to plan across the different urban authorities which collectively make up the Katowice agglomeration, despite the fact that key social, economic and ecological processes unite these political areas. The SKAP was supported by the formation of the Union for the Sustainable Development of the Cities of the Katowice Agglomeration. This union brought the constituent urban authorities of the agglomeration – Bedzin, Bytom, Chorzow, Czeladz, Dabrowa Gornicza, Jaworno, Katowice, Myslowice, Pietary Slakie, Ruda Slaska, Siemianowice Slaskie, Sosnowiec and Zarbre[1] – together, so that social, economic and environmental planning could be more effectively co-ordinated.

Beyond greater policy co-ordination, the SKAP embodies practices which are clearly differentiated from the previous socialist plans. First, the SKAP is based upon the active participation of urban residents in the development of its plans and schemes. This public openness is in stark contrast to the socialist programmes which were forced on Katowice under socialism, but is also based on the belief that in order to be sustainable any post-socialist urban development scheme must win the active consent of the people that it will most directly effect. A further key aspect of the SKAP has been the production and dispersion of data relating to socio-environmental conditions in the city (known as the *Environmental Profile*). One of the characteristics of socialist urbanism in Katowice was the lack of available data on health and environmental pollution. Furthermore, even when this data was available, it was routinely withheld by the state and shrouded in secrecy. Through the use of different inspection agencies and local authorities, the SKAP has sought to maximize the production of data on socio-environmental conditions in the city. This has meant that progress in improving the city can be more carefully monitored and people living in the city can have a clearer picture of what the quality of life in Katowice is actually like relative to other urban centres. The democratization of socio-environmental data in cities like Katowice has been one of the characteristic shifts associated with the rise of post-socialism in Europe (see Pavlínek and Pickles 2000: chapter 7). A final aspect of the SKAP which is worthy of note, relates to its programme of industrial landscape restructuring. In order to simultaneously improve the quality of life in the city and repair the damaged industrial landscapes produced under socialism, the SKAP has initiated a series of programmes which are devoted to the sustainable reuse of key industrial sites throughout the agglomeration. Related programmes range from the re-cultivation of waste disposal areas to the conversion of heavy industry plants to cleaner post-industrial uses.[2]

At one level there have already been clear benefits from the SKAP. With improved waste treatment facilities in the city there has been a decline in water pollution events. At the same time the trans-boundary pollution emanating from Katowice has also been greatly reduced. As with many incidents of reduced environmental pollution in the post-socialist world, however, it is difficult to know whether it is sustainable development policies or simply the decline of heavy industries which have made a difference in Katowice. What is clear is that short-term environmental gains made in cities like Katowice are likely to be threatened in the longer term as the mass consumption economics of neo-liberalism brings more waste and car-based pollution to Eastern Europe (see Baker 2000: 162).

What the case of Katowice illustrates are the ways in which issues of sustainability were crucial both to the crisis of socialism and to the emerging socio-political systems and practices associated with post-socialist society. This case study also illustrates the danger of simply seeing sustainability as an imported model of social development which has been implemented in post-socialist communities after the collapse of state socialism. The example of Katowice illustrates that many of the current practices and ideals associated with sustainability in Poland were emerging long before the socialist systems collapsed. In recognizing this complex history of post-socialist transition, it becomes easier to discern how post-socialist states have developed models of sustainability which are often different in their style and emphasis than those promoted elsewhere in the world. In this context, it is very interesting to reflect upon Jaroslaw Sarul's (the State Secretary for the Environment in the Republic of Poland in 2000) description of Poland's particular brand of sustainability. Sarul argues that emerging sustainable development policies in Poland are akin to developing a *sense of solidarity with nature* (Sarul 2000). The notion of solidarity with nature is intriguing not just because it echoes Poland's radical political past, but because of what it tells us of emerging Polish visions of sustainability. It appears that just as the success of Solidarity as a political movement was based upon the mobilization of unprecedented political support and co-operation for workers' issues, the future social and economic success of the country depends upon achieving a partnership of respect with nature. Just as the struggle against socialism was based on political togetherness, it appears that the struggle against contemporary social and environmental problems in Poland is seen as being best resolved through a unified response to the interrelated social, economic and ecological challenges facing the country. The emerging link between sustainability and notions of social and environmental unity in Poland is an interesting expression of the inherent geography of sustainability, but it also reveals the different cultural and political values which are being attached to notions of sustainability in the post-socialist world. These values – often drawing on important historical reference points and phrases – appear crucial to giving the struggle for a sustainable future particular meaning and significance.

Exploring sustainability in the post-socialist world II: sustainability and the Russian experience

The final example of the relationship between sustainability and post-socialist transition I want to consider is that of Russia. Russia and the former Soviet Union provide a particularly interesting insight into the contested relations between sustainability and post-socialism. In part this is because as the hub of the Soviet-style system of European state-monopoly socialism, Russia has been more resistant than many countries to the onset of post-socialism. Russia is, however, of particular interest in the context of this book because of the ways in which the principles of sustainability have merged alongside pre-existing socio-cultural and scientific traditions to generate a relatively distinct geography of sustainability. Much of the discussion conveyed in this section is based on Jonathan Oldfield's extensive analysis of sustainable development in Russia (see Oldfield 1999, 2000, 2001; Oldfield and Shaw 2002).

Sequencing sustainable development in Russia

As the ideological, administrative and military hub of European socialism, Russia's relationship with the principles and ideals of sustainability is particularly instructive. The Soviet Union-led boycott of the United Nations Conference on the Human Environment in Stockholm in 1972 meant that the USSR was partially isolated from many of the formative discussions which would eventually result in the principles we associate with sustainable development today.[3] Despite this, however, the Russian Federation has actually been a firm supporter of the principles of sustainable development, adopting a range of United Nations-led initiatives and agreements pertaining to sustainability (Oldfield 2001: 96).

Russia's formal engagement with the principles of sustainable development really began in 1992 when it took a full and active part in the United Nations Conference on Environment and Development in Rio de Janeiro. After agreeing to the main policy goals of the Rio summit, the principles of sustainable development gradually started to influence domestic policy and development strategies in Russia (ibid.: 96). A key moment in Russia's adoption of sustainable development was the delivery of a presidential decree committing the Russian Federation to the production of a state plan for sustainable development (ibid.). This presidential decree described sustainable development as a model for post-socialist society through which the economic, social and environmental needs of the country could be reconciled. Ultimately this decree led to the production of a Russian government plan for sustainable development policies from 1994–1995 (ibid.). In 1996, this initial vision for sustainable development was supplemented by an additional presidential decree concerning *the transition to sustainable development* in Russia (ibid.: 97). What is particularly interesting to me about this decree are the ways in which it links transition with sustainable development. This decree was devised to guide post-socialist transition in Russia along a more sustainable trajectory. In the very production of this decree then we see the problem with accounts of transition which see it simply as a move from socialism to unregulated, free-market capitalism. While the pressures of unfettered capitalist development are undeniably strong in countries like Russia, it is clear that the principles of sustainable development are being promoted as a strategy for developing a more social justice and an ecologically tolerable future.

In order to support policy strategies and plans since 1992, the Russian government has been committed to passing various forms of environmental laws designed to ensure sustainable forms of post-socialist development (ibid.: 98). These laws have been applied to a range of different environmental contexts, including inland waters (internal seas) and the use of the continental shelf and subsurface resources, and have been supported by the passing of a *Water Code*.[4] This explosion of environmental legislation is not untypical of what has been happening in other post-socialist states. In their analysis of the Czech Republic, for example, Pavlínek and Pickles describe a *legislative revolution* around the environment emerging in the country (2000: 195–202). Pavlínek and Pickles claim that post-socialist transitions have been characterized by a keen legislative interest in environmental issues. According to Oldfield's analysis, however, one of the most intriguing and original elements of the Russian national vision of sustainable development is the way that it envisages the temporal movement from a socialist to a sustainable society (2001: 103). While conventional wisdom on post-socialist transition has argued that the drive towards sustainable development has often taken second place to the priority of structural economic transitions, in Russia the order of priority is seen slightly differently (at least in official terms). While acknowledging the need to address the interrelated problems of Russia's society, environment and economy together, Russia's strategy clearly sees a *sequencing* of activities and priorities as being important (ibid.). In this context, the national strategy describes a sequence of stages through which the nation's transition to sustainability should be pursued (ibid.: 103–104). Significantly, the first stage described in the 1996 presidential decree is tackling prevailing social and environmental problems in those areas which have experienced the most adverse effects of socialism (ibid.: 103). Only once these serious socio-environmental problems have been resolved, it is argued, can the next stage of full economic reform, envisaged in the national plan, be pursued (ibid.). The third and final stage (or order of priority) described in the decree, is to develop a deeper awareness of the role of Russia's socio-environmental relations within the wider global environment and society (ibid.: 104). In the face of unremitting pressure for economic reform, it is questionable whether the envisaged sequencing of sustainability in Russia has proved to be achievable. What the case of Russia does illustrate, however, are the complex and diverse models of sustainability which are emerging in the post-socialist world.

Sustainable development the Russian way

Given the rapid adoption of sustainable development by the Russian Federation following the collapse of socialism and the ensuing Rio Earth Summit, Russia's relationship with sustainable development could be seen as relatively straightforward. In line with conventional accounts of sustainable development in the post-socialist world, it appears that once free from the shackles of socialist rule and ideology, a particular brand of UN-sponsored sustainable development has spread from the global community into Russian society. The work of Oldfield, however, suggests that far from simply adopting a pre-existing model of sustainable development, a distinctive Russian way of doing sustainability has been emerging over the last fifteen years. Oldfield traces the roots of sustainability in the Russian Federation to ideas and philosophies which emerged in the socialist and pre-socialist periods.

In relation to pre-socialist traditions, Oldfield and Shaw (2002: 395) draw parallels between current discourses of Russian sustainability and much older notions of *sobornost*. The notion of *sobornost* originated within traditional forms of agricultural community, within which harmonious socio-ecological relations were pursued through the establishment of close ties to the land and an intimate knowledge of how ecological systems operate and survive (ibid.). The idea of understanding nature's ways and local environmental processes is a crucial theme within Russia's emerging plans and strategies for developing a more sustainable society. Related to the notion of *sobornost* is the later notion of the *noosphere*, which became popular among scientists during the Soviet era (see Oldfield 2001: 104). Developed by the Russian scientist V.I. Vernadsky, the idea of the noo-sphere essentially refers to a 'sphere of wisdom' within which society will restructure its value systems and modes of knowledge to ensure a greater harmony with the natural world (ibid.: 104). Interestingly, in the work of Vernadsky, the noosphere is the last of a series of stages of social devel-opment, or evolution, and represents a point at which rational and utili-tarian views of the natural world are finally replaced by a fuller spiritual understanding of humanity's place in nature (ibid.). While Vernadsky's idea of a noosphere is actually not that far removed from many existing deep-green philosophical traditions in the West, it has clearly had a signifi-cant impact on sustainable development in post-socialist Russia. This influence can been seen in both the sequencing, or staging of sustainable development, as well as the strong association which is being made in Russia between the achievement of sustainable development and the spir-itual growth of the nation. This form of sustainability is clearly different from the more rational, business-oriented versions of sustainable develop-ment which appear to be predominant within many Western governmental plans. In this sense, it is important to recognize that while post-socialist states have clearly been influenced by pre-existing ideas of sustainable development, there are lessons which the rest of the world can also take from the forms of sustainable development which are now being imple-mented in the post-socialist world.

Beyond the emergence of a geographically distinctive discourse of sustainable development in Russia, it remains difficult to assess the actual impact of these new policies in practice. Despite the enactment of impres-sive new environmental laws, it is unclear whether the Russian govern-ment has the capacity to effectively implement these policies. The danger of course remains that as the pressures of global free-market capitalism take a deeper hold on Russian society, sustainable development policies could be downplayed in favour of unregulated economic development (see Baker's 2000 work on environmental policies in former East Germany). At this stage, then, while the sustainable intentions of the Russian govern-ment are clear, the pursuit of sustainable development in Russia remains a fragile and potentially vulnerable process.

Summary

In this chapter we have considered the emergence of sustainability in the post-socialist world. The historical coincidence of the end of socialism in Europe and the Rio Earth Summit (at which the international political community ratified sustainable development as a key principle for future socio-economic policies) has resulted in sustainable development being enthusiastically embraced by many post-socialist nations. Consequently, throughout post-socialist Europe and in Russia it is possible to discern the incorporation of sustainable development into a range of key strategies for national development, new legislation and associated systems of socio-environmental rights. This chapter's analysis of the rapid uptake of sustainability in the post-socialist world has, however, revealed two crucial things. First, that sustainable development has played a complex and at times ambiguous role in post-socialist transitions. At one level, it has provided an alternative vision of the future to the unregulated free-market capitalist system routinely pursued under post-socialist governments. At times, however, the calls for sustainable development have been drowned out by the supposedly more urgent needs for economic reform. Second, sustainable development has been interpreted differently and adapted to a greater or lesser extent in different post-socialist states. Consequently, whether it is Poland's vision of solidarity with nature, or Russia's sequential vision of the noosphere, sustainability has not simply been transferred to the post-socialist world. Sustainability has clearly emerged from within the upheavals of post-socialist transitions as both an international policy goal and the product of more locally based political and scientific traditions. What our reflections on the sustainable society in the post-socialist world thus show is that geography matters when we are trying to understand and interpret the emerging forms of sustainability which are currently in existence.

Suggested reading

There are a number of good introductory texts to the geographies of post-socialist societies and associated socio-political transitions. See for example: Bradshaw, M. and Stenning, A. (eds) (2004) *East Central Europe and the Former Soviet Union: The Post-Socialist States*. There are, however, far fewer detailed studies of the links between post-socialism and sustainability. Perhaps the best extended study is Pavlínek, P. and Pickles, J. (2000) *Environmental Transitions: Transformations and Ecological Defence in Central and Eastern Europe*, although this volume focuses more on environmental issues than sustainability. For a more detailed analysis of the role of sustainability within post-socialist society, however, see the collected work of Jonathan Oldfield: Oldfield, J.D. and Shaw, D.J.B. (2002) 'Revisiting sustainable development: Russian cultural and scientific traditions and the concept of sustainable development', in *Area*, 34: 391–400; Oldfield, J.D. (2001) 'Russia, systemic transformation and the concept of sustainable development', *Environmental Politics*, 10: 94–110; Oldfield, J. (1999) 'Socio-economic change and the environment – Moscow city case study', *Geographical Journal*, 165: 222–231.

Suggested websites

For more on the sustainable urban development in Katowice: http://www.unhabitat.org/programmes/sustainablecities/katowice.asp

More about the Solidarity movement can be found at: http://www.bellaonline.com/articles/art34963.asp

4 THE POLLUTION OF POVERTY

Sustainability in the developing world

Introduction

At the beginning of their excellent book *Just Sustainabilities: Development in an Unequal World* (2003), Agyeman *et al.* establish a new definition for sustainable development. To replace the now well-rehearsed version of sustainability established by the World Commission on Environment and Development as: 'development that meets the needs of the present without compromising the ability of future generations to meet their own needs' (WCED 1987: 43), Agyeman *et al.* instead interpret sustainable development as 'the need to ensure a better quality of life for all, now and into the future, in a just and equitable manner, whilst living within the limits of supporting ecosystems' (Agyeman *et al.* 2003: 2). While I freely admit that this new definition may lack some of the simplicity and directness of its predecessor, I think that it embraces some critical insights into the nature of the sustainable society which are regularly overlooked. The point is that within the WCED's definition of sustainable development, the idea of achieving the needs of present generation remains an unqualified assumption. I say unqualified because this definition fails to place the concept of need within a global context. As soon as we begin to interpret need within a global context, we become aware of great variations within what different people define as needs and of the great disparity which exists in the extent to which different people have their needs met. At one level, it is possible to calculate with a reasonable degree of accuracy what the basic needs of all humans are. These would include the intake of certain levels of nutrition, access to shelter and health care and the maintenance of a clean environment within which to live. Beyond these needs, many of us define our needs in very different ways. Many people living in Britain claim that they need a car, without which their lifestyle of commuting to work and living in the countryside would not be possible. To others living in some of the most impoverished parts of the world, the idea of 'needing' a car would seem absurd. The reason for this disparity in the definition of need unquestionably stems from the fact that in many parts of the world the basic needs which humans require simply to live are not being adequately met. United Nations figures reveal that nearly eight hundred million people in the world today do not get enough food to eat, while approximately five hundred million are designated by the United Nations as being *chronically malnourished.*[1] According to the *World Development Report 2000–2001*, 2.8 billion people live on less than two (US) dollars a day, while 1.2 billion have to survive on less than one (US) dollar a day (World Bank 2000). Even more worrying perhaps from UN figures is that 1.2 billion people live without access to safe drinking water.[2]

What I am trying to convey is that social, economic and environmental needs are not internationally homogeneous, but vary greatly from country to country and place to place. Furthermore, as UN figures reveal, certain people's, place's and country's needs are much more urgent than others. Accepting the idea that need is a relative concept – that is to say that its definition varies from place to place and person to person – it is important to realize that if sustainability is concerned with the balancing of social, economic and environmental needs, then this balancing process, and the relative priority given to social, economic and environmental issues, will obviously vary from place to place and between one person and another. This chapter explores how, in the context of the extreme forms of poverty and social disadvantage which many people now face in less economically developed countries, a brand of sustainability which is partic- ularly concerned with social sustainability – or simple human survival – has emerged in many parts of the world. In making this statement, I am well aware of the dangers of geographical generalization that it entails. I do not want to suggest that the alleviation of poverty and social depriva- tion effectively captures the essence of all of the many and varied sustain- able development schemes currently being pursued in less economically development countries (hereafter LEDCs). I do, however, want to empha- size how the problems and historical causes of poverty in LEDCs provide a persistent context within which related sustainable development strate- gies should be understood and interpreted.

This chapter is made up of three main sections. The first section explores the colonial and imperial histories that have contributed to develop- ment problems throughout LEDCs, before considering how contemporary social and scientific research within LEDCs has revealed a harmful cycle emerging between environmental destruction and the production of poverty. The second section considers early post-war attempts to instigate new development strategies in LEDCs through a discussion of the so-called *Green Revolution* in Asia. In the context of the failings of the Green Revolution, the final section then explores the emergence of sustainable development in LEDCs, focusing specifically on Kenya and the contemporary tensions which surround forest management in certain parts of the country.

Pollution of poverty: the challenge of development in less economically developed countries

Some readers may be surprised to see that I refer to LEDCs rather than perhaps the better known notions of the 'developing' of 'third world'. In focusing on LEDCs I am, however, not referring to a different set of places to those which could also be classified as a third world, but to a different way of understanding these diverse and complex spaces. While the problems associated with the terms 'developing' or 'third' world are now well rehearsed, I would like to convey briefly why I do not use them. At one level I dislike the idea of a 'developing world', because it suggests a set of places which are somehow cut off from the rest of global society. Theories of international development have consistently shown that in order to understand the 'developing world' – or the presence of a world which is economically and socially disadvantaged relative to other places – we must have some knowledge of the colonial and imperial links which have existed between the developing and developed world over the last five hundred years. At another level contemporary research has begun to question whether 'developing world' countries actually have enough in common to be classified together at a geographical level. The socio-economic disparity which exists between supposedly kindred developing world states has been emphasized by United Nations figures which show an increasing disparity between the poor and poorest of nations of the 'developing world' – a group of countries which are experiencing extreme forms of poverty even when compared to other developing world nations. In a related sense, many now recognize that the neat geographical division between LEDCs and more economically developed countries (hereafter MEDCs), actually hides the overlaps which exist between these two categories. There are for example pockets of 'first world' privilege and affluence which exist within many LEDCs, while enclaves of 'third world' poverty are found within many MEDCs (regions of extreme poverty and underdevelopment in the MEDCs are sometimes referred to as the 'Fourth World', Potter *et al.* 2004: 27). This situation clearly makes any clean geographical distinction between a developing and developed world hard to justify.

It is in this context that I prefer the notion of LEDCs, because it draws our attention to the particular developmental needs of certain countries not 'worlds'. In essence the idea of LEDCs emphasizes the importance of a careful geographical analysis of the developmental needs of different states and regions which, while recognizing their common need for forms of development, does not ignore their different political, economic and cultural traditions. The remainder of this section explores the historical roots of poverty and underdevelopment in LEDCs; how the pressures of poverty and the need for poverty alleviation have hindered historical programmes of environmental management in LEDCs; and finally how recent interpretations of sustainable development within LEDCs have enabled a clearer link to be established between the double social and environmental 'pollutions' of poverty.

Problems of poverty: its components and its history

In order to understand the links between sustainability and poverty in LEDCs, it is necessary to take a brief step back in time. Many analysts argue that the relative social and economic disadvantage being experienced by LEDCs is not merely a contemporary phenomena, but can be traced back into a much longer history of colonial exploitation and imperial expansion. Despite their importance within the history of LEDCs, much confusion surrounds the terms colonization and imperialism (see Potter *et al.* 2004: 50). Collectively the terms colonialism and imperialism are usually used interchangeably to refer to the spread of Western European political, economic and military influence throughout the world. Despite this tendency, however, colonialism and imperialism do actually have quite distinct and separate meanings. According to Michael Watts (1995: 75), colonialism involves '[t]he establishment and maintenance of rule, for an extended period of time, by a sovereign power over a subordinate and alien people that is separate from the ruling power'. When colonialism is associated with the political and military practices of colonization, Watts states that it 'involves the physical settlement of people (i.e. settlers) from the centre to the colonial periphery' (p. 75). Crucially, Watts does not see colonialism as being synonymous with imperialism, but instead states that 'colonialism is a variant of imperialism' (p. 75). Watts understands imperialism as a much broader set of processes than colonialism, involving an 'unequal territorial relationship among states, based on subordination and domination, and typically associated with [. . .] the emergence of monopolies and trans-national enterprises' (p. 75) (see also Harvey 2003a). If the physical acts of political settlement and occupation associated with colonialism are only one, admittedly extreme, form of imperialism, we are immediately forced to consider the different ways in which processes of subordination and domination between countries can become manifest. The point is that while the age of European colonial empire may be disappearing (particularly with the growing wave of national independencies which have replaced Spanish and Portuguese domination in South America; French, British, Dutch, Italian and Belgium control in Africa; and French, Dutch and British influence in Asia), the legacies of colonialism still remain, while the forces of imperialism (or neo-imperialism) continue to shape the world around us. Certain writers have identified the manifestation of new forms of colonial and imperial power in the global age: Hardt and Negri (2000) have famously described the emergence of a new 'Empire' of political control and imperialism within the hidden and seemingly innocent processes of globalization; in a similar way, the geographer Derek Gregory (2004) has described a *Colonial Present* emerging in the *post-September the 11th era*, as US-led military activities in Afghanistan and Iraq recreate the colonial tensions of the past.

The role of colonialism and imperialism in shaping our contemporary world is a complex story. The histories of European colonial power and imperial interventions must be understood, however, if contemporary attempts to create more sustainable societies in LEDCs are to have any success (however that may be measured). The history of Western imperial interventions within what we now term LEDCs began in the sixteenth century with the emergence of something called *mercantile colonialism*. Mercantile colonialism was essentially a period of time when European powers engaged with a range of nations through a complex system of commerce and trade. While the buying and selling of goods did not involve the political occupation of space, which we classically associate with colonialism, it did involve European trading companies locating in distant trading places and regulating the emerging systems of international capitalist trade (Ogborn, 2002). While certain forms of international trade did benefit non-European communities in some ways (particularly in terms of the growing availability of Western goods), trade relations were often constructed in a way which deliberately privileged the European trader (Potter *et al.* 2004: chapter 2). Consequently, European traders would often use the political power of their home nations to force through trade concessions which benefited them, but disadvantaged their trading partners. In certain instances there is also evidence that such traders would engage in more direct acts of economic plunder and resource theft.

In addition to the imperialistic trade relations associated with early mercantile capitalism, another form of imperialism started to emerge. This form of imperialism involved states taking political control of former trading regions and claiming their assets and peoples for their own needs. These patterns of direct territorial acquisition, military occupation and political control are those which we classically associate with colonialism (see Potter 2004: 62–68; Scott 2004). If mercantile capitalism had disadvantaged non-European states because of the use of trade concessions and unfair trading practices, the colonial period saw a more overt form of exploitation taking place. This process of colonial exploitation involved the seizure of lands and assets from local people, which were then exploited in order to supply European markets. In order to utilize these new lands and resources, Western colonizers would often employ local labourers who would be worked mercilessly for little or no pay. Many theorists of colonial history claim that the expansion of European economic activity into new colonial areas provided the foundations for rapid industrial expansion and wealth creation in Western Europe, while European states deliberately suppressed domestic economic expansion within the colonies. This period of direct European colonial expansion was not, however, just about economic policy. This time was also synonymous with the promotion of Western cultural, political and religious values throughout the colonies (see Said 1995). In this context, European colonizers often portrayed indigenous beliefs and practices as 'backward' and 'unsophisticated', while European models of development were praised for their civility and rationality. Significantly, the historical imposition of Western models of development appears to continue to inform the suspicions which many LEDCs feel towards contemporary doctrines of sustainable development (Dresner 2002: chapter 3). Beyond the economic and cultural legacies of this era, Western states also irrevocably altered the political landscapes of their colonies. In many instances the administrative governments set up by colonizers provided the nuclei for the state apparatuses of newly emerging independent states in the immediate post-colonial era. At the same time, the territorial markers of colonial empires (which often cut across pre-existing cultural and ethnic spaces) were utilized by post-colonial states to mark out their areas of political and economic sovereignty (Potter *et al.* 2004: 73–79).

One of the most troubling legacies of Western colonialism in LEDCs was environmental destruction and degradation. The economic success of colonial endeavours often depended on the rampant exploitation of ecological resources which routinely displayed scant regard to the long-term effects such practices would have on the environment. Running parallel to these forms of environmental exploitation was a form of environmental imperialism which has both informed and inhibited contemporary strands of sustainable development. In the middle to late colonial period (nineteenth and twentieth centuries) Adams charts the emerging desire of many colonial states to actively conserve and protect certain environmental areas within LEDCs (see Adams 2001: 23–34). This desire to protect colonized environments was connected to an increasing realization within the industrial world that the environments of Africa, Asia and Latin America embodied a kind of unspoilt, pre-industrial Eden (for a more detailed discussion of the history of the Edenic myth see Merchant 2004). Such a belief led to the imposition of game reserves, nature conservation areas and national parks in many LEDCs. While offering important forms of environmental protection, such conservation areas have been interpreted by many living in LEDCs as environmental playgrounds for wealthy Western tourists pursuing safari or adventure holidays. It has also been argued that the environmental protectionist discourses of the colonial era have placed unfair restrictions on the economic activities of LEDCs, which may need to exploit protected environmental resources for economic advantage. It is in the context of this legacy of environmental imperialism that many leaders of LEDCs have questioned the motives behind the sustainable development agenda and opposed it as simply another form of imposed Western environmental ideology.

As I have already intimated earlier in this chapter, perhaps the most telling legacy of European imperialism in LEDCs is that of poverty. During early discussions of sustainability at an international level, the problems associated with poverty in LEDCs were often referred to in relation to the phrase the 'pollution of poverty' (Adams 2001: 56). The use of *pollution* as a metaphor for talking about poverty in LEDCs is important for three main reasons. First, it is clear that the environmentally infused concept of *pollution* was used to illustrate that while Western governments appeared preoccupied with environmental protection, many in LEDCs remained devoted to addressing the social damage caused by poverty. Second, the term pollution – understood as harmful or unwanted goods – served to emphasize the serious forms of damage and social harm which poverty causes throughout LEDCs. Third, and finally, the notion of pollution also served to emphasize the unusual concentration of poverty within certain LEDCs – it is after all the concentration of unwanted/harmful substances which transforms them into pollution. The United Nations currently defines poverty in the following terms:

> Denial of choices and opportunities most basic to human development, reflected in a short life, lack of basic education, lack of material means, exclusion, and a lack of freedom and dignity. Human poverty is multidimensional and people-centered, focusing on the quality of human life rather than on material possessions.
>
> (UN Human Development Report 1997, Glossary of
> Poverty and Human Development, quoted in Lead 2004)

The crucial thing to recognize within this definition of poverty is the emphasis which it places upon the multifaceted nature of poverty. Poverty is not just a question of a lack of income (although access to financial resources plays a crucial role in determining people's ability to access housing, proper nutrition, health care and education). The pollution of poverty is consequently expressed in a variety of ways in LEDCs, ranging from poor health, a lack of sanitation, exploitative working conditions and poor housing. Often these different facets of poverty are mutually re-inforcing and tend to create a poverty trap from which it is increasingly difficult for people to escape.

If you accept a multidimensional definition of poverty, it obviously becomes difficult to accurately assess when and where poverty exists. In order to assist in the exercise, however, the World Bank has developed a measure of poverty which is called a *poverty line*. According to the World Bank, a poverty line works on the basis that those deemed to be above the line are not in poverty, while those below the line are designated as poverty stricken (Hayat 2004). While crude in many ways, this measure of poverty does provide us with some useful insights into the nature and concentra-tion of poverty in LEDCs. While different poverty lines have been set for different world regions (in order to allow for international comparison), the general cut-off point for individual or family income is one (US) dollar a day. Consequently, any family or individual with an income of less than one (US) dollar a day is designated by the World Bank as being in poverty (ibid.). As mentioned previously current figures show that 1.2 billion people live on less that one (US) dollar a day, with 2.8 billion people surviving on less than two (US) dollars every day (ibid.). While these figures are disturbing in their own right, it is the realization that the inter-national poverty which they reflect is concentrated in LEDCs which is perhaps most concerning. The concentration of poverty in LEDCs can be seen in a variety of ways. At one level this concentration of poverty can be understood in relation to food security, with 160 million pre-school chil-dren in LEDCs being designated as underweight.[3] At another level this intensity of poverty can be conceived of in relation to certain dimensions of human health, with 93 per cent of HIV/AIDS sufferers currently living in what the United Nations designates as 'developing countries'.[4]

As we have seen, the causes of poverty are complex, but two things clearly emerge from the analysis of poverty in LEDCs presented so far. First, that with such a high concentration of human poverty existing in LEDCs, any discussion of sustainability in the context of these commun-ities is going to focus on issues of human sustainability – or put another way, of survival. Second, it is clear that while poverty in LEDCs is in certain instances being perpetuated by poor government structures in the post-colonial era, the roots of LEDCs' poverty lie firmly in the systems of unfair and exploitative socio-economic exchange established under European imperialism. While traditional forms of imperialism may be in decline, many now observe a new era of neo-imperialism emerging within which the developed world is once again exploiting weaker LEDCs through unfair trading systems, market protectionism and even military intervention (see Harvey 2003a). In this context it becomes clear that strategies designed to address issues of sustainability in LEDCs must tackle the problematic history of exploitation created between the developed world and LEDCs.

Environmental pollution of poverty

While the eradication of the social suffering associated with poverty has been a crucial goal of sustainable development, sustainability is also premised upon a concern for the harm which poverty can inflict on the environment. In many ways arguing for a link between poverty and environmental exploitation seems paradoxical and counterintuitive. It was after all economic success and rapid industrial growth which generated the first forms of international political concern over global environmental pollution. Increasingly, however, policy makers have been confronted with the reality that while unregulated economic growth can be harmful to the environment, poverty can be disastrous. The link between social poverty and environmental destruction is aptly expressed by Jonathan Porritt when he observes:

> Poverty is one of the greatest threats to the environment today. It is poverty that drives people to overgraze, to cut down trees, to adapt ecologically damaging shortcuts and lifestyles, to have larger families than they would otherwise choose, to flee from rural areas into already over-burdened cities – in short to consume the very seed corn on which the future depends, in order to stay alive today.
>
> (Porritt 1992: 35)

Porritt's point is that when people are in poverty they are essentially forced to exploit their surrounding environment as much as they possibly can, simply to stay alive. In this context, it is clear that a farmer who is struggling to feed their family will do everything they can in the short term to ensure that their land produces enough food to keep their family alive. In this sense, it is clear that poverty forces people to think about short-term, immediate needs and not to plan for a sustainable future. To a poverty-stricken farmer, for example, the idea of leaving land fallow for a number of seasons to enable it to recover its productive capacity through nutrient renewal would seem ludicrous, particularly if it was unlikely that they would live to see the benefits of this newly fertile land. Of course in the long term, the overexploitation of land will simply make it more and more difficult for the farmer to provide for their family, as their land gradually yields less produce. In the case of this fictitious (but all too real) farmer then it is possible to discern the emergence of a poverty trap, or a trap of *unsustainability* (see Figure 4.1). The trap of unsustainability is premised upon the principle that once in poverty, people are forced to exploit their surrounding environment in such a way that its long-term productive capacity is compromised. The reduced productive capacity of the environment in turn leads to a larger-scale economic downturn, involving decreased land values and investment, which only serves to perpetuate social poverty. Of course the reaffirmation of social poverty again results in the need to exploit the environment in an unsustainable way, and so the dreadful cycle continues.

An interesting insight into the links between social poverty and environmental degradation can be found in a number of writings on soil erosion and desertification, which emerged in the 1980s (see Blaikie 1985; Blaikie and Brookfield 1987; Grainger 1982). Severe periods of drought, soil erosion and famine during the 1970s and 1980s led to a renewed endeavour in the 1980s to uncover what caused soil erosion and its concomitant social and ecological problems. The writings of Piers Blaikie are in many ways emblematic of this tradition. Drawing on research in India, Nepal and Africa, Blaikie sought to explore what caused soil erosion in these particular places. At the heart of Blaikie's innovative analysis was a desire to

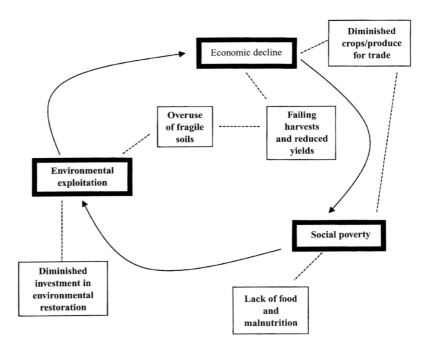

Figure 4.1 The trap of unsustainability

challenge the conventional accounts of soil erosion which had emerged within environmental science. Many scientific analyses of soil erosion attributed the phenomenon to physical environmental factors such as climatic intervention, soil quality and slope gradients. According to Blaikie (1985), however, these scientific approaches to soil erosion failed to recognize that the degradation of soil was just as much a political and economic process as it was a factor of the physical environment. In this context Blaikie states:

> [s]oil erosion is a political-economic issue [. . .] of course soil erosion is an *environmental* process as well. Within a geological perspective is goes on with or without human agencies [. . .] It becomes a social issue when this deterioration (or removal of the soil) is recognised and some sort of action taken.
>
> (Blaikie 1985: 1–2; original emphasis)

By recognizing the links which exist between soil erosion and political economic processes, Blaikie draws our attention to both the types of social relations which generate erosion in the first place, and to the relative levels of social action (and inaction) which are mobilised to stop erosion and enhance soil conservations.

Although Blaikie reveals the multiple and varied social and environmental factors which cause soil erosion, he recognizes that rural inequalities and disadvantage are often at the heart of soil erosion problems. In this context Blaikie (1985: 3) states that 'soil erosion is a symptom of underdevelopment, and it reinforces that condition'. In the context of soil erosion among small-scale agriculturalists, Blaikie recognized that poverty and disadvantage within LEDCs is the outcome of three processes: (1) the legacies of colonial trade and exploitation; (2) the emergence of new patterns of international exploitation in LEDCs often carried out by large multinational corporations; (3) the exploitative relationships established between peasant farmers and landowners within certain rural areas, which resulted in those who own land taking excessive amounts of farmers' surplus products as rent. Crucially in the context of this chapter, the link between poverty, soil degradation and the trap of unsustainability described by Blaikie is also experienced in a wide range of other ecological and economic contexts, including forestry and fishing. The work of Blaikie and others writing within the political ecology tradition highlights the links which exist between the legacies of imperial and colonial exploitation, social poverty and environmental destruction in LEDCs. Blaikie's analysis of soil erosion also reveals that any attempt to restructure environmental relations in LEDCs must address socio-economic issues of reform as well as ecological ones.

From Green Revolution to sustainable development

As the previous section has illustrated, a fundamental paradox is faced by many LEDCs. At one level increasing incidences of social poverty and expanding populations in LEDCs mean that people need their respective environments to provide more energy, more food and more economic resources than they have ever done before. At the same time, however, they are faced with the harsh reality of increasingly degraded and unproductive environments and dwindling environmental resources. The question of how to take more from the environment, while degrading the ecological resources less, lay at the heart of post-war debates over development in many LEDCs.

The West knows best: technological imports and the Green Revolution

In order to meet increasing social demands for food and agricultural produce in LEDCs, many development agencies and governments argued that it was necessary to instigate more efficient, scientifically informed systems of environmental utilization. At the heart of this philosophy was the argument that if environments in LEDCs were reaching the natural limits of their potential utilization, then artificial ways needed to be developed in order to increase the productive capacity of the environment. Specifically in terms of agriculture, this idea was realized in the bewildering array of new pesticides, artificial fertilizers, new agricultural technologies and miracle crops, which became synonymous with the *Green Revolution* (see Adams 2001: 300–301; Yapa 1996). The Green Revolution, as it was euphemistically entitled, involved the export of Western agricultural practices and techniques to LEDCs. By using modern agricultural technologies it was anticipated that food harvests and yields could be improved and crop losses due to drought and insect/fungal attacks diminished.

The Green Revolution was led and instigated by a number of Western governments and associated economic institutions. According to Shiva (1991a), many of the initiatives which were developed under the banner of the Green Revolution were support by the US government and wealthy American institutions. It appears that the US government were concerned that newly independent states in Asia, Africa and South America could follow China's socialist systems of agricultural reform and abandon its own model of liberal capitalist development. In order to ensure that Western models of agricultural reform were established throughout LEDCs, the US government offered funding and subsidies for Green Revolution policies (ibid.). At the same time, the wealthy American-based institutions of Ford and the Rockefeller Foundation came together to form the *Centro*

International Agriculture Tropical (CIAT) (based in Columbia) and the *International Institute for Tropical Agricultural* (IITA) (based in Nigeria) to support Green Revolution policies in South America and Africa respectively (see Shiva 1991a). The basis for the US-led vision of a Western agricultural revolution in LEDCs was fourfold: the use of new high-yielding varieties (HYVs) of crops; the application of non-organic, artificial pesticides and fertilizers; the introduction of new forms of mechanized farm technology; and the development of new irrigation schemes (Shiva 1991b). It was anticipated that the integrated application of these 'modern' agricultural technologies could simultaneously solve the problems of food supply and continued environmental destruction in LEDCs. In this sense, the Green Revolution was predicated on the belief that with the appropriate use of technologies, more could be taken from the environment, with less environmental damage being caused.

The socio-ecological problems caused by the Green Revolution as it spread throughout much of Africa, Asia and South America during the 1960s and 1970s have been the subject of a range of critical commentaries and books. My personal favourite is Vandana Shiva's marvellous *The Violence of the Green Revolution* (1991b). In this provocative and disturbing volume, Shiva unpacks the fundamental contradictions and unsustainable nature of the Green Revolution (see Box 4.1). In environmental terms, Shiva argues that the spread of HYV monocrops and inorganic pesticides and fertilizers has caused widespread environmental violence in LEDCs. At one level this environmental harm has been expressed in the loss of biodiversity. In place of varied crop types, the widespread use of monoculture wheat and rice eroded the natural biodiversity upon which agricultural communities had historically relied (for more on the consequences of these processes see Shiva's discussion of *Biopiracy*, Shiva 1998). The spread of intensive forms of monoculture crops has, according to Shiva, contributed to increasing patterns of soil erosion and the greater vulnerability of harvests from attacks by pests and fungi (1991a). Related to the loss of ecological diversity, the use and transmission of artificial fertilizers and pesticides has been linked to the proliferation of water pollution throughout LEDCs. In social terms, Shiva also indicates that the Green Revolution has failed many farmers. Because of the emphasis which the Green Revolution placed on Western technologies and chemicals, many smallholding farmers became dependent on purchased inputs to maintain their farms (Shiva 1991a, 1993). When these farmers fell on hard times, however, it became difficult for them to repair new farm technologies, or afford the crop varieties and fertilizers upon which they depended. In this context, it has been noticed that the Green Revolution was only beneficial in the long term to the large agro-industrial operations in LEDCs who were able to support technologically intensive farming practices. Meanwhile poorer,

Box 4.1 Vandana Shiva on the *Violence of the Green Revolution*: 'soil bandits' in the Punjab

Vandana Shiva is one of the leading international writers on environmental politics, women's rights and agriculture. She is the director of the Research Foundation for Science, Technology and Natural Resource Policy, and in 1993 she won the Right Livelihood Award (known as the Alternative Nobel Prize). While famous for writing on a range of issues, Vandana Shiva is perhaps best known for her critique of the effects of Western policies on the agricultural politics and development of LEDCs. It is in this context that Vandana Shiva became one of the most outspoken critics of the Green Revolution. The first lines of a famous piece which Shiva wrote for *The Ecologist* reveals her feelings about the Green Revolution:

> The Green Revolution has been a failure. It has led to reduced genetic diversity, increased vulnerability to pests, soil erosion, water shortages, reduced soil fertility and availability of nutritious food crops for the local population, the displacement of vast numbers of small farmers from their land, rural impoverishment and increased tensions and conflicts.
>
> (Shiva 1991a: 57)

Perhaps what is most surprising about this rather unambiguous critique of the Green Revolution is that it is based upon farming reform in the Punjab region of India – the home to supposedly the most successful example of what the Green Revolution has achieved. Through a careful analysis of agricultural policy and practice in the Punjab, Shiva illustrates that the Green Revolution was primarily benefiting the agrochemical industry, not small-scale farmers. She argued that the Green Revolution was converting the farmers in the Punjab into unwilling *soil bandits*, who were damaging the agricultural environments upon which they depended. According to Shiva, the exploitation of the agricultural peasantry in the Punjab (who are predominantly Sikh) has contributed to the emerging support which the Punjab nationalist movement is receiving (Shiva 1991a). By attacking the Green Revolution through the example of the Punjab, Vandana Shiva struck at the very heart of the scientific principles and ideals which the revolution was based upon.

Key reading: Shiva, V. (1991a) 'The Green Revolution in Punjab', *The Ecologist*, 21: 57–60; Shiva, V. (1991b) *The Violence of the Green Revolution: Third World Agriculture, Ecology and Politics*.

more marginal farmers suffered and were unable to sustain the initial benefits which the Green Revolution brought them (Shiva 1991b).

The uneven impact of the Green Revolution on large- and small-scale farmers respectively has generated something of a paradox. This paradox is predicated on the fact that while food production in many LEDCs is increasing, it is not being matched by a proportional decline in undernourishment and malnutrition. The Food and Agriculture Organization consequently state that while food production has increased dramatically in LEDCs (from 2110 calories per capita in 1969, to 2680 calories per capita in 1999), the number of people classified as undernourished has only fallen from 956 to 777 million in the equivalent time period (see Masanganise and Swaminathan 2004). The reasons for this paradox are complex, but it is clearly related to the fact that increases in food production are largely occurring in the large, often export-oriented agro-industrial companies of LEDCs, while poorer subsistence farmers still find it difficult to produce the food they require to stay alive (Shiva 1991a, b). In the context of the social and environmental failures of the Green Revolution, some have claimed that as a political and economic act, the revolution was nothing more than a form of neo-imperialism. Yapa (1996) argues that by both underestimating the value of local farmers' practices and ecological knowledge and ignoring the reproductive capacities of nature, the Green Revolution embodied the imposition of a supposedly rational, Western model of development onto supposedly 'backward' socio-economic systems. What then emerged was an agricultural system which was able to produce cheap foodstuffs for Western markets, and which was dependent on Western companies for the regular supply of overpriced chemical products and technologies (see also Adams 2001: 300–301).

Rise of sustainable development in LEDCs

In many ways the emergence of sustainable development policies in LEDCs reflects a response to the failings of externally imposed programmes like the Green Revolution. While it is impossible to provide a full account of all of the nuances associated with sustainable development policies in LEDCs, by breaking down policy initiatives into local, national and international arenas, it becomes easier to understand the complexities of sustainability in LEDCs. (For a more detailed review of sustainable development policies in the LEDCs see Adams 2001; Elliot 1999; and Harrison 1993.)

Beginning then at a local scale, one thing which is noticeable about emerging forms of sustainable development in LEDCs is the value which is placed upon the local itself (see also Chapter 8). By *valuing the local*, I am referring to the ways in which sustainable development prioritizes the use of local knowledge and resources within its various social, economic and cultural policy areas. The value of the local is expressed very clearly in an agricultural context in the Agenda 21 document which was prepared at the Rio Earth Summit:

> The rural household, indigenous people and their communities, and the family farmer, a substantial number of whom are women, have been the stewards of much of the Earth's resources [. . .] A farmer-centred approach is the key to the attainment of sustainability in both developed [. . .] and developing countries and many of the programme areas in Agenda 21 address this objective [. . .] The decentralization of decision-making towards local and community organizations is the key in changing people's behaviour and implementing sustainable farming strategies. This programme area deals with activities which can contribute to this end.
>
> (Agenda 21, Chapter 32, UNCED 1992)

As this quote reveals, an important principle of sustainable development is the idea of decentralized decision making. By the decentralization of decision making I am referring to the process whereby local people are empowered to develop locally feasible strategies of development.

The decentralization of decision making is important for two reasons. First, because it marks a clear historical distinction from the eras of imperialism colonialism and neo-imperialism when people living in LEDCs were told how to develop. By giving people the power and resources to shape their own futures, those advocating sustainable development argue it is much more likely that people will pursue more sustainable lifestyles – even if it is only out of enlightened self-interest. Second, the decentralization of decision making to a local level is important because it recognizes the valuable forms of ecological/environmental knowledge which local communities often hold (see Raffles 2002: chapter 1). Again quoting from Agenda 21:

> Indigenous people and their communities have an historical relationship with their lands and are generally descendants of the original inhabitants of such lands [. . .] They have developed over many generations a holistic traditional scientific knowledge of their lands, natural resources and environment. Indigenous people and their communities shall enjoy the full measure of human rights and fundamental freedoms without hindrance or discrimination. Their ability to participate fully in sustainable development practices on their lands has tended to be limited as a result of factors of an economic, social and historical nature.
>
> (Agenda 21, Chapter 26, UNCED 1992)

While there is a sense of local community idealism evident in this quote, it does emphasize that local knowledge should be valued and not simply rejected as backward and unscientific within Westernized programmes of socio-economic reform (what counts as local knowledge is, however, often difficult to ascertain, and many social scientists now prefer to think in terms of *networks of knowledge*, rather than local, national or international forms, see Latour 1993). The emphasis which is placed on localization within sustainable development thinking, extends beyond decision making to incorporate the use of local resources and technologies within development schemes. One of the most obvious reasons why the Green Revolution was unsustainable was because it was based on the importation of expensive equipment and technologies. Once these new technologies failed, poor farmers were unable to replace them and the pattern of agricultural production they had adopted was undermined. By utilizing and adapting local practices and knowledge, sustainable development seeks to implement projects and modes of development which are quite literally locally sustainable. Despite the progressive intentions of *local empowerment* associated with sustainable development, however, significant neo-imperial barriers continue to exist to sustainability being achieved at a local level. In a new era of biotechnology, genetic patents and intellectual property rights legislation are serving to disempower farmers in LEDCs as multinational agricultural conglomerates control the use and development of new biotechnologies (see Shiva 1998). It appears that just as the technological issues surrounding local empowerment are being addressed in LEDCs, the potential basis for technological disempowerment has shifted from the human to the genetic level.

In the context of instigating locally sensitive development schemes, policies for sustainable development have also required the creation of a very different role for governments at a national scale. While the principles of sustainable development envisage an important role for national governments in developing distinctive and locally sensitive brands of sustainable development, they also emphasize that state governments should not support the national imposition of *top-down* development schemes as they often did in colonial times (although for a more complex account of the role of states in development project see Robbins 2000). In this context, national governments have two crucial roles within the sustainable development process in LEDCs. First, they offer the necessary financial and administrative support for locally sustainable development strategies to emerge. Second, they should provide strategic guidance to larger-scale sustainable development policy areas such as energy production, population control and national resource management. In these two contexts, state policies for sustainable development in LEDCs are perhaps best thought of as frameworks within which sustainability can be co-ordinated and harmonized between different local communities. In this way, Agenda 21 understands national governments not so much as implementation agencies for sustainable development, but as *enabling states*, providing the institutional, legal and financial support for sustainable development policies to emerge. At the Rio Earth Summit it was consequently agreed that:

> Each country should aim to complete, as soon as practicable, if possible by 1994, a review of capacity – and capability-building requirements for devising national sustainable development strategies, including those for generating and implementing its own Agenda 21 action programme. By 1997, the Secretary-General of the United Nations should submit to the General Assembly a report on achievement of improved policies, coordination systems and procedures for strengthening the implementation of technical cooperation programmes for sustainable development, as well as on additional measures required to strengthen such cooperation. That report should be prepared on the basis of information provided by countries, international organizations, environment and development institutions, donor agencies and non-governmental partners. Activities [should include] Build[ing] a national consensus and formulating capacity-building strategies for implementing Agenda 21.
>
> (Agenda 21, Para 37.4, UNCED 1992)

In this context, states have a crucial role in co-ordinating sustainable development policies in LEDCs, and also of ensuring that such policies reflect the values and priorities of that country, not those of an external power.

Finally, the goal of more sustainable forms of development in LEDCs is dependent upon action taken at an international scale. The Brundtland Report (WCED 1987) argued that achieving sustainable development in LEDCs is not just a technological issue, involving the development of new locally sensitive social, economic and ecological practices, but that it has to tackle the broader international legacies of imperialism and colonialism in LEDCs. In particular, the Brundtland Report highlighted the need for trade reform and debt relief if sustainable development is going to be achieved in LEDCs (WCED 1987: chapter 3). In terms of international trade, the Brundtland Report argued that MEDCs were using their economic power to protect their own markets from LEDCs' export goods, while flooding the domestic markets of LEDCs with their own surplus products. Through the use of stringent tariffs and various market protectionism mechanisms, MEDCs have consistently defended their economies from relatively cheap imports from the developing world. This of course prevents LEDCs from earning valuable forms of foreign exchange which could bolster their economies (ibid.). A good example of this unfair system of trade is provided by the case of sugar (see WCED 1987: 82). Sugar cane is a vital agricultural commodity in many LEDCs. The Brundtland Report recognized that many MEDCs have provided subsidies and trade protection for beet sugar production (which they are often more self-sufficient in), which has in turn had an adverse effect on the international sugar cane market. According to the *World Development Report* (World Bank 1986), MEDCs' policies on sugar trade cost LEDCs US$7.4 billion in lost trade earnings in 1983 (WCED 1987: 82). It was, according to the Brundtland Report, these broader economic structures of trade exploitation that made it difficult to eradicate social poverty and achieve sustainable development in the LEDCs (ibid.: 78–83).

While this book does not analyse trade relations in depth, it is important at this point to point out the broader significance of trade to sustainability debates. Many claim (see above) that unfair trading relations lie at the heart of the poverty and socio-economic unsustainability associated with many LEDCs. Others argue that heightened levels of global trade are environmentally damaging – resulting in the wasteful use of energy and resources. At one and the same time, however, there are many politicians and leading economic experts who claim that more open trading systems are the key to forging more sustainable futures in LEDCs. What is indisputable is that trade is central to the economic, social and environmental issues associated with sustainability. Perhaps the best indication of the positive effects which trade reform can have on sustainable development in LEDCs is provided by the success of the Fair Trade movement. By offering a fair and consistent price for goods produced in LEDCs, the Fair Trade movement is supporting environmental care and community building in many states, while also securing the economic stability of many agricultural groups.

In relation to international debt, the Brundtland Report stipulated that sustainable development would be almost impossible to achieve in certain countries unless they received assistance with their crippling repayments (WCED 1987: 73–75). During the post-Second World War period, many LEDCs borrowed money from international banks and organizations (like the World Bank), in order to facilitate their development plans. While the repayments of these debts was not problematic during times of economic growth and relative prosperity, when recession and economic crisis hit the international economy during the 1970s, it became increasingly difficult for LEDCs to service their debts (ibid.: 73). The Jubilee Debt Campaign estimates that 52 LEDCs currently owe $300 billion in debt and that African countries alone spend approximately $40 million everyday repaying their existing loans (Jubilee Debt Campaign 2002). As the Brundtland

Report emphasizes, this colossal burden of debt is socially and environmentally unsustainable. At a social level debt is unsustainable because it often results in national governments cutting back public spending on health, education and public utility maintenance in order to repay the interest accumulated on loans. The constant repayment of international debts also places great pressure on environmental resources in LEDCs, as governments look to exploit new environmental assets as a basis for creating new trade surpluses, which can be used to service historical debts. In the context of the unsustainable nature of international debt in LEDCs, the Brundtland Report, and other international agencies and UN bodies, have been calling for new international policies to deal with debt. These policies often incorporate debt-relief programmes, which either involve the forgiveness of outstanding debt balances or new agreements being reached which allow international debts to be paid back over longer periods of time (WCED 1987: 75).

As this section has shown, just as distinctive forms of sustainability are emerging in the MEDCs and the post-socialist world, in LEDCs it is possible to discern a different brand of sustainable development. Policies for sustainability in LEDCs have sought to address the forms of social, economic and environmental disadvantage which many living in LEDCs suffer. Just as convincing industrial leaders in the West that sustainable development is good business, a key part of achieving sustainability in LEDCs has been allowing sustainable development to address local issues in local ways. This focus on local empowerment has been vital in ensuring that sustainable development is not simply another facet of neo-colonial control and management as practised by the West on LEDCs. Despite isolating the key characteristics of sustainable development in LEDCs, however, our discussions have so far remained fairly abstract. The remainder of this chapter consequently explores the contested implementation of sustainable development in Kenya. This case study illustrates both the value and limitations of sustainable development policies as they are currently being delivered in LEDCs.

Sustainable forestry, women's rights and the Green Belt Movement in Kenya

Local sustainability and the Green Belt Movement

Our story begins in the foothills of Mount Kenya and with the work of Wangari Maathai. Wangari Maathai is a prominent environmentalist and women's rights activist in Kenya. She has been a long-term opponent of the Kenyan government's support for forest clearances at the base of mount

Kenya in order to make property available for various economic purposes. This protest is part of a longer history of social and environmental opposition instigated by Wangari Maathai. During the 1970s Wangari Maathai become concerned about the erosion of forest resources and woodlands in Kenya. Kenya's extensive woodlands had already been severely depleted

---

Box 4.2 The story of Wangari Maathai

The personal story of Wangari Maathai – founder of the *Kenyan Green Belt Movement* – is a fascinating one. Born in 1940, Wangari Maathai travelled to America to study biology and gained a PhD in veterinary medicine. On returning to Kenya she taught veterinary medicine and eventually become the head of her university department. In 1977 she established the Green Belt Movement in response to the destruction of Kenya's forest resources and the socio-environmental problems this was creating. In 1987 Wangari Maathai formed the first Green Party in Kenya in order to give her environmental values a more formal political voice. As a result of environmental activism and green politics, Maathai become unpopular among Kenya's political elites and in the early 1990s she was arrested by the Daniel Arap Moi government. On gaining her freedom, Wangari Maathai stood for president in 1997 (her nomination was, however, nullified because of a technical political rule) before winning a place in parliament in 2002 (Selva 2004).

In 2004, Wangari Maathai's amazing political career and activism was rewarded when she received the Nobel Peace Prize. The prize was awarded on the basis of Maathai's revolutionary work in protecting the *life environment* of many living in Kenya and her tireless campaigns for women's rights in Africa. The idea of the *life environment* which the Nobel committee commented upon, is of course a very sustainable notion. It is a concept which seems to automatically recognize the role of environments in the support of both ecological and social life, and the need to preserve environmental resources while ensuring social needs are also met. Wangari Maathai is the first African woman to win the Nobel Peace Prize and the first person to receive the award for environmental activities (Selva 2004).

Key reading: Maathai, W. (2004) *The Green Belt Movement: Sharing the Approach and the Experience*; Selva, M. (2004) 'Queen of the Greens', *The Independent*, 9 October: 38–39.

Also visit: http://www.greenbeltmovement.org/index.php

during the colonial period, when wood was used to support imperial economic expansion. More recently, Kenya's forests have been cleared by the state to make way for new developments, or by farmers to bring new land into productive use. It is now estimated that 75 per cent of Kenya's forests have been cleared over the last 150 years.[5]

Even during the 1970s, when the concept of sustainable development was only an emerging embryonic notion, Wangari Maathai recognized that Kenya's treatment of its forests was unsustainable. Maathai argued that it was unsustainable for three reasons: (1) because it was generating environmental damage, both in terms of the loss of biodiversity and the escalation of soil erosion events which followed forest clearances; (2) the impacts which forest clearances were having on the local communities which depended on woodlands for their fuel wood and food; (3) forest clearances often resulted in the alienation of local communities from the environment as outside property developers and state farmers took control of once communally used local environmental resources. In light of the interrelated socio-environmental problems associated with forest clearances, in 1977 Maathai formed the *Kenyan Green Belt Movement* (hereafter GBM). The GBM is a non-governmental agency which is devoted to developing a sustainable response to the destruction of Kenya's forest resources. Originally established as part of the *National Council of the Women of Kenya*, the GBM is devoted to forest regeneration and biodiversity protection which is based upon community tree-planting and forest management schemes (see Maathai 2004).

While at one level the activities of Maathai's Green Belt Movement make it seem just like any other environmental activist group, a closer look at its activities reveals links with notions of sustainability. The links between the GMB and the practices and ideas of sustainable development are really clear at two levels: first in relation to the ways in which the GBM understands forests; and second in the context of the particular local programmes which the GBM runs. One of the most important achievements of the GBM has been the development of a more holistic, or integrated, view of Kenyan forests. Essentially, the GBM has challenged the idea that forests should be conceived of purely as economic resources – as they have historically been by the state and colonial powers. Instead the GBM argues that forests contain crucial environmental and social resources, in addition to serving important economic functions (see Maathai 2004). In relationship to the environment, the GBM has consistently stressed the importance of forest ecosystems as a crucial context within which biodiversity can flourish and be sustained. In a social context, the GBM has also highlighted the role of forests as an important source of fuel wood and foodstuffs upon which a range of local communities depend for their everyday survival. By highlighting the role of forests in supporting numerous ecological and social communities, the GBM has developed a powerful critique of narrowly conceived clearances of Kenya's woodlands.

Beyond the sustainable conceptualization of forests as social, economic and environmental spaces, the GBM has also developed a range of sustainable practices and schemes. In keeping with classical sustainable development philosophy, the GBM has fostered a range of locally based and community-oriented projects for forest regeneration. The GBM's main project areas incorporate: environmental conservation; civic and environmental education; Green Belt Safaris; and capacity building for women's groups (ibid.). In relation to environmental conservation, the GBM's main project has centred upon community tree-planting. Since the late 1970s, the GBM has been supporting local groups in the formation of tree nurseries. These nurseries involve local community members (particularly women) in the planting and rearing of trees.[6] The seeds produced by these trees are then given to local farmers in an attempt to encourage them to devote more of their land to forest cover. Those working within the tree nurseries are paid by the GBM and receive extra payments for the seeds which they produce (ibid.). Many of these seeds are now being used to support the GBM's attempts to extend new tree-planting programmes on public lands. According to the GBM, the main driving force behind this tree-planting and forest conservation strategy is a belief that in order to sustain life, a nation should have at least 10 per cent of its land area covered by woodlands.[7] At the present time it is estimated by the GBM that only 2 per cent of Kenya's land area is forested – making the country unsustainable in terms of its oxygen/carbon dioxide balance; its ability to produce key food and fuel types; and its natural soil fertility (Green Belt Movement 2005). By fostering the planting of new trees in community nurseries, on private agricultural property and on public land, the GBM hopes to redress this socio-ecological imbalance in Kenya. By 1997, twenty years after the inception of the GBM, twenty million trees had been planted by the organization (Green Belt Movement 2005). In addition to its tree-planting programmes, the GBM has also been active in promoting new forms of civic and environmental education. Through its Learning Centre in Nairobi, the GBM runs seminars, school lessons and lecture series which are designed to raise public awareness of environmental issues and give people the skills which they need to develop their own sustainable development schemes. So far it is estimated that 10,000 people have received education and training as part of the GBM programme.[8]

The GBM has also been active in developing a series of community-based safari schemes. Once again in keeping with the principles of sustainable development, these GBM Safaris are not like the mass tourist events typically promoted by the state and large travel companies, but rather combine some tourist activities with community-based work and educational sessions. Typically, community safaris last between five and seven days and involve conventional safari activities and trips with time set aside for visitors to actually work in the community tree nurseries. Visitors are hosted by local families, who in turn benefit from the tourist income which the safaris generate. While obviously generating an important source of local revenue, community safaris also serve to remind local people of the value of preserving their local environments and its associated wildlife, because these are often precisely the assets which visitors travel to Kenya to experience. A final programme initiative for which the GBM has now become well known is its *Women for Change* project. This initiative involves young women, who are often employed within tree nursery schemes, and seeks to guide them in making complex decisions concerning their sexual and reproductive health as well as promoting healthy eating practices. An important part of the Women for Change project is the empowerment of women. This empowerment really operates a two levels: first in relation to the income which they receive for working on forest projects; and second with regard to giving women the knowledge they need to live healthier and safer lives. In the context of an often male-dominated society, and the increasing dangers associated with HIV/AIDS in Africa, the Women for Change initiative reflects an interesting example of how social and environmental sustainability can be integrated. This project is also a powerful reminder of the important links which exist between sustainability and the consolidation of basic human rights.

The combined initiatives which have developed as part of the GMB reflect many of the principles of sustainable development we have already discussed in this book. At one level, by understanding woodlands as simultaneously social, economic and environmental resources, the GMB has reaffirmed the integrated socio-ecological doctrines of sustainable development. Perhaps what makes the GBM's projects most sustainable, however, are the emphases which they place upon local capacity building, community knowledge and social empowerment. By emphasizing the role of local communities in forest projects, the GBM has ensured that their initiatives are not perceived as imposed programmes of development, such as those experienced in the colonial era. By giving local communities the chance to participate in and to develop their own forest management programmes, the GBM has ensured that such schemes have the best chance of being sustainable, or lasting, programmes of socio-environmental development. As with much sustainable development thinking, the GBM has realized that any programme of socio-environmental reform is much more likely to be a long-term success if it is acceptable to local people, who can then internalize it into their everyday practices and values.

National sustainability and poverty reduction strategies

While the Green Belt Movement has historically been an activist organ-ization, which has operated at a distance (and often in direct antagonism with the state), since 2002 things have begun to change. In 2002 the *National Rainbow Coalition* came to power in Kenya and defeated the Daniel Arap Moi administration. The National Rainbow Coalition came to power on the promise of socio-economic and environmental reform (Selva 2004). In relation to its environmental promises, the National Rainbow Coalition made Wangari Maathai its minister for the environment. Many hoped that with this appointment it would be possible to develop a more sustainable set of social, economic and environmental policies at a national level in Kenya. As we have previously mentioned, nation states are believed to have crucial roles in setting the broader political and economic contexts within which sustainable development can be achieved across different local areas. In light of this responsibility, in June 2003 the National Rainbow Coalition launched a *poverty reduction strategy* (with an associated *economic recovery strategy*), as a national framework within which to foster sustainable development. Poverty reduction strategies are economic initiatives which are being promoted by the World Bank and International Monetary Fund (IMF). Poverty reduction strategies involve individual nations devising strategies of political and economic reform and poverty reduction which they submit to the World Bank and International Monetary Fund for assessment (this process is normally referred to as a joint staff assessment because it involves staff from both the World Bank and IMF). If ratified by the joint staff of the World Bank and IMF, applicant countries become eligible for international aid to assist their socio-economic restructuring and associated attempts to tackle poverty.

If we take a closer look at Kenya's poverty reduction strategy, it is possible to discern precisely what the World Bank and IMF believe is necessary to ensure social and economic stability in the country. To begin with it is interesting to note that the production of Kenya's poverty reduc-tion strategy involved extensive consultation with various civic and private interest groups in Kenya. This consultation was carried out as part of the development of a national poverty eradication plan and was designed to ensure that any national poverty reduction strategy reflected the priorities and beliefs of the people of Kenya. On the basis of this consultation exer-cise, the Kenyan government produced an economic recovery strategy which was to provide the basis for its poverty reduction objectives. This economic recovery strategy applied to the agricultural, manufacturing and tourist sectors of Kenya's economy and emphasized the importance of export promotion and financial sector reform (International Monetary Fund

2003). Ultimately the economic recovery strategy recognized that the only way of tackling Kenya's endemic poverty was through macro-economic reform (ibid.). In this context, the government essentially recognized that while local projects like those developed by the Green Belt Movement made a real difference to local communities, they could not address poverty across the whole of Kenya's economy. What was required then, it was argued, was a broader strategy of economic growth, the proceeds of which would be strategically targeted at poverty alleviation programmes (ibid.).

Kenya's poverty reduction strategy suggests that the key to achieving sustainable economic growth (and thus addressing national poverty) are four key economic objectives: (1) containing inflation below 5 per cent; (2) reducing the public sector wage bill; (3) establishing a clear system of public sector privatization; and (4) reducing economic debt and deficits (International Monetary Fund 2003: 5–6). While all of these objectives clearly address important macro-economic issues, they are problematic for a number of reasons. First, these proposed reforms clearly reflect a set of Western, market-based economic values. At the heart of these values lies the belief that the key to solving social poverty is to free the market from the inefficient and often corrupt grip of big government. This neo-liberal economic belief system is essentially what lies behind the proposed reform of the public sector proposed in Kenya, and the opening up of Kenya's national economy to international trade. Of course, as with many neo-liberal programmes of economic growth, such beliefs appear to neglect the potentially harmful and exploitative economic consequences which inter-national trade can bring to LEDCs, while assuming that the economic benefits of national growth will filter down (through poverty alleviation programmes) to the poorest in society. Consistently, economic analyses have shown that free trade is not always beneficial to developing econ-omies and the economic benefits of national growth do not trickle down to those most in need, without the presence of a strong welfare state bureau-cracy. In this context, it is clear that contemporary poverty reduction strategies, like that currently being developed in Kenya, are in danger of failing in the same way that the World Bank and International Monetary Fund's *structural adjustment programmes* did in the 1990s. Based upon similar patterns of neo-liberal economic reform (i.e. the liberalization of the market; tight control of public expenditure), structural adjustment programmes are blamed by many for contributing to serious forms of economic collapse in many LEDCs.

While reading Kenya's poverty reduction strategy, it becomes quickly apparent that despite occasional reference to sustainable economic growth, there is no connection made between Kenya's programmes of economic reform and poverty reduction and its national environmental policies. Is it to be assumed that economic growth and poverty alleviation will produce no new pressure on national environmental resources? While, as we have seen, the tackling of poverty should not necessarily be seen as incompatible with environmental protection and enhancement, it is clear that a sustainable national strategy for poverty reduction should make some reference to how poverty alleviation is aligned to sustainable environmental management. The absence of reference to sustainable environmental management within Kenya's poverty reduction strategy means that it remains unclear as to what happens when economic growth and poverty reduction threaten environmental resources and amenities. How are such conflicts between economic plans, social need and environmental protection to be resolved? Such uncertainty has already led to the continuance of forest clearances in Kenya to make way for luxury housing and to Wangari Maathai threatening to resign from the new government in protest (Selva 2004: 39).

From the brief case study of Kenya, it is clear that just as in MEDCs and the post-socialist world, particular geographical circumstances in LEDCs are influencing the ways in which economic, social and environmental priorities are being addressed in emerging sustainable development schemes. In the case of LEDCs, it is clear that the urgent need to eradicate serious forms of social poverty is creating a danger, at least at macro-economic level, that environmental priorities are marginalized. Furthermore, by invoking neo-liberal strategies of economic reform (which are supported by the World Bank and International Monetary Fund) there is a risk that many LEDCs, like Kenya, are going to recreate the same exploitative international economic relations which led to the concentration of international poverty in their nations in the first place.

Summary

In this chapter we have considered how and why the alleviation of social poverty has become a key priority of policies for sustainabilities in LEDCs. By looking at the level and extent of poverty within LEDCs, we have seen that defining the term 'need' – a term which is routinely used within discussions of sustainable development – often takes a very different form in LEDCs than it does in other, more affluent parts of the world. In order to understand why poverty remains such a prominent feature of life in LEDCs, this chapter has briefly explored the complex histories of imperial and colonial exploitation which generated this situation. In light of this historical analysis, this chapter has emphasized that any attempt to generate more sustainable patterns of development in LEDCs must recognize the continuing legacies of exploitation which exist in these countries today.

The persistent concentration of poverty in LEDCs has led many to argue that in the future the reduction of poverty, through economic growth, could cause major environmental damage in the poorest parts of the world. In this chapter we have seen, however, that it is not just economic growth which threatens the environment, but that in many cases poverty can prove to be more dangerous to the environment than affluence. In the context of the simultaneous need for poverty alleviation and environmental protection in LEDCs, this chapter has explored how the Green Revolution offered one potential solution to the problems of impoverished nations. Based upon the use of Western technologies and scientific advances, the Green Revolution sought to make environments more productive through the application of fertilizers, pesticides and high-yielding crop varieties. As an expensive form of imposed agricultural reform, this chapter revealed that the Green Revolution only tended to help large agro-industrial businesses, and while increasing aggregate food production, it failed to significantly address the persistent problems of malnutrition and food distribution in LEDCs. In the context of the failures of the Green Revolution, this chapter moved on to consider how sustainable development emerged during the 1980s as an alternative strategy for social development and environmental protection in LEDCs. Based upon an appreciation of the links which connect social and economic development, with the use and protection of the environment, we have seen how sustainable development has been implemented through a series of local, national and international actions and policies. By focusing upon the case of Kenya, we have seen the application of local modes of sustainable development through the example of Wangari Maathai's Green Belt Movement. Based upon empowering local communities to manage and protect their local woodlands as a source of food, fuel and economic income, the Green Belt Movement embodies a quintessential form of sustainable development. A brief analysis of Kenya's new national strategies for economic reform and poverty reduction has, however, revealed the potential difficulties which surround effectively implementing sustainable development policies at a national level in LEDCs. With its emphasis on market liberalization and privatization, Kenya's poverty reduction strategy reminds us that there is a real danger that while local sustainable development schemes like those developed by the Green Belt Movement continue to thrive, they may be undermined by the re-creation of the same (neo-)imperial modes of economic exploitation which lie at the heart of many LEDCs' social, economic and environmental problems today.

Suggested reading

The most comprehensive and accessible account of sustainability in LEDCs is provided by Adams, W.M. (2001) *Green Development: Environment and Sustainability in the Third World* (second edition). For an excellent critique of sustainable development policies in LEDCs see O'Riordan, T. (1989) 'Politics, practice and the new environmentalism', in D. Gregory and R. Walford (eds) *Horizons in Human Geography*, chapter 6.2.

Excellent analyses of the impacts of the Green Revolution in India are provided by Shiva, V. (1991a) 'The Green Revolution in the Punjab', *The Ecologist*, 21: 57–60; Shiva, V. (1991b) *The Violence of the Green Revolution: Third World Agriculture, Ecology and Politics*.

For a detailed insight into the Kenyan Green Belt Movement see Wangari Maathai's own book – Maathai, W. (2004) *The Green Belt Movement: Sharing the Approach and the Experience*.

Suggested websites

The latest United Nations figures on poverty and malnourishment in less economically developed countries can be found at: http://www.undp.org/teams/english/facts.htm

For more information on international debt see the Jubilee Debt Campaign home page at: http://www.jubileedebtcampaign.org.uk/

For a more detailed account of the activities of the Kenyan Green Belt Movement and Wangari Maathai go to: http://www.greenbeltmovement.org/index.php

PART II
Scales of
sustainability

5 SUSTAINABILITY IN A GLOBAL ERA

Introduction

The year 1968 is synonymous with social revolution and an emerging
desire to try and reshape the world in which we live. Whether it was in
relation to the peace protests surrounding the Vietnam war, student demonstrations on the streets of Paris, or urban race riots in the United States of
America, 1968 was a time when people were questioning the moral and
political assumptions upon which (at least Western) society was based (for
a fascinating discussion of the relationship between the year 1968 and
changing geographical ways of understanding the world see Watts 2001).
On Christmas Eve 1968, however, images sent back from the Apollo 8
space mission would provide a new challenge to the ways in which
humanity understood the world in which we live. The now famous Apollo
8 images were the first live pictures of the planet Earth transmitted from
space. In addition to showing the beautiful fragility of the Earth, set against
the inky darkness of space, the Apollo 8 mission also captured pictures of
the Earth rising over the surface of the moon. In many ways it is difficult
to overestimate the profound impact these pictures of the Earth had on the
collective political consciousness of global society. In an attempt to try and
capture the profound geo-historic importance of these Earthly images,
rather than proving a commentary to support the pictures they were
beaming back, the crew of Apollo 8 decided to read from the book of
Genesis (and in particular the story of creation). Quite what people from
non-Judaeo-Christian religious backgrounds made of the crew of Apollo
8's choice of reading is difficult to assess, but it is clear that the use of an
evocative piece of scripture, recounting the creation of the Universe,
reflected the emotive potential of these new pictures.

The pictures which reached Earth on Christmas Eve 1968 provided the
iconic imagery for two socio-political forces which were set to dominate
the final quarter of the twentieth century. In the first instance the image
of the planet Earth has become the dominant motif associated with the

complex mix of social, cultural, political and economic forces which are collectively referred to as globalization (see Cosgrove 2001). In this context, the idea of a single planet moving in isolation through space, confirmed by the Apollo mission, served to underline the global connections which underpin human existence and life (Pickles 2004: 78–79). In a related, but very different context, the image of the Earth from space also became a crucial inspiration and symbolic icon of the environmental movement (Ingold 1993). To many in the fledgling green movement of the late 1960s, the image of the Earth from space served to indicate two things: first the idea that the Earth is a self-contained, environmentally integrated system; and second that the Earth is a finite and fragile resource which should be cared for and protected. This chapter considers the relationship between these emerging understandings of the planet as a globalizing system and debates surrounding sustainability. In relation to these aims, we will see that the emergence of notions of sustainability in an era of renewed planetary consciousness was no mere coincidence. Consequently, in this chapter we chart how a growing awareness of the global dynamics of environmental change and socio-economic interaction (see Liverman 1999) have been central factors within the formation and implementation of the principles of sustainability.

This chapter begins with a critical investigation of the concept of globalization and how latest geographical work is challenging many of the underlying assumptions associated with the analysis of global phenomena (see also Chapter 1). Through the use of two case studies, the remainder of this chapter then explores the relationship between the processes of globalization and sustainability. In the first instance we consider the emergence of global political responses to sustainable development issues. Focusing in particular on the issue of climate change, this section considers the importance of developing international policies for sustainable development which cut across national boundaries, but also reveals the political difficulties associated with this process. In the second case study we consider the emergence of sub-global but supra-national strategies of sustainable development. Drawing on the example of environmental management in the Barents Sea, analysis considers the importance of developing environmental management strategies which harmonize different national policy regimes.

Geographies of globalization: on conceptual hangovers and the scale debate

Unpacking globalization

Globalization is one of those words which appears to be used with far more frequency than understanding. As a word, globalization entered the popular social and political lexicon in the early 1960s, but it now appears clear that the processes of globalization have been operating for many centuries (see Hirst and Thompson 1996). When trying to come to terms with what globalization actually is, it is helpful to begin by dispelling any notion that it refers to one thing or one set of processes. In this context Peter Dicken argues that globalization is actually a set of *overlapping discourses* which are used to understand and transform the world (2002: 315). Supporting this view, David Harvey argues that it is possible to break globalization down into three interrelated categories: a process; a condition; and a political project (2000: 54). As a condition, globalization is perhaps best thought of as a mode of being within which people, economies and cultures are becoming increasingly integrated and connected. In this context, global forms of integration are based upon new modes of communication and information technology (Castells 2000); new types of social mobility and transport (Urry 2000); and the functional integration of economic systems (Harvey 1989). As a political project, globalization can be understood as a strategy of global economic and political expansion (pursued largely by Western nations and corporations) to open up new markets and sources of labour under the ideologies of free trade and neoliberalism. In this context, the discourse of globalization is often described by politicians as a basis for transforming and improving the world, but can equally be maligned as a cause of economic decline within a given state (see Swyngedouw 1997a on the scalar discourses of globalization). Finally, when understood as a process (Harvey's preferred understanding of the term) globalization involves the *geographical reorganization of capitalism* (Harvey 2000: 57).

Understanding globalization as the *geographical reorganization of capitalism* has important implications. At one level globalization has been understood as the creation of a borderless world of infinitely mobile bodies and processes (see Ohmae 1990). In this context, globalization has gradually become synonymous with the erosion of the sovereign power and influence of nation states (see Chapter 1) – as containers of political and economic power – as global processes effortlessly transgress and permeate their borders (for an excellent critical analysis of this perspective see Brenner 2004). Harvey's assertion that globalization involves the spatial reconfiguration of capitalism does, however, challenge such notions. Rather than seeing globalization as the end product in a linear movement from territorial states to non-territorial economies, Harvey interprets globalization as part of the ongoing spatial de- and re-territorialization of capitalism. In this context, rather than being an end of geography, or the power of space, many geographers are now interpreting globalization as the context within which a new array of spatial forms and modes of existence are being created which offer fresh opportunities for economic dominance and political resistance (see for example Bridge and Wood 2005; Routledge 2003; Sharp *et al.* 2000; Sparke 2004).

Recent work by geographers has consequently sought to understand how the flows and processes associated with globalization are *reshaping* not *replacing* spatial existence. In the context of this research agenda, an increasingly large number of geographers have focused upon the issues of scale and how globalization has been rescaling the geographical categories in and through which we organize the world (Brenner 1999; Jones 1998; MacLeod and Goodwin 1999; Marston 2000; Smith 1992, 1993; Swyngedouw 1997a, b; Whitehead 2003b). Historically, of course, geography has a long association with the study of scales. Neighbourhoods, regions and nations are scales around which our world has been organized and have provided the loci for a range of different geographical modes of enquiry. The contemporary *scale debate* does, however, differ from geographers' conventional approach to scale. It differs in the way in which it questions the extent to which scales can actually be interpreted as fixed geographical layers of existence. In the broad context of the spatial transformations associated with globalization, those analysing scales have started to question understandings of scales as rigid hierarchies of spatial organization, within which purportedly *natural* scales like the home, neighbourhood, city, region and nation are all neatly nested together in a kind of *Russian doll* framework (see Whitehead 2003b: 285). Instead, scalar analysts claim that globalization is seeing the disintegration of certain scales and the formation of new scalar frameworks of political, social and economic life. These new scales include urban innovation districts, city regions and supra-national coalitions (see Brenner 2004). This debate has been famously described as *the hangover after the party which was globalization.*[1] On these terms, the scale debate is perhaps best thought of as an attempt to replace the discourse of a de-territorial, almost weightless existence associated with globalization studies, with a more careful excavation of the new geographies which are emerging as part of the contemporary transformations associated with global development.

The scale debate has recently been criticized from a range of different perspectives. Some argue, for example, that it produces a simplified vision of the world based upon a series of hierarchically organized, arbitrarily conceived levels of existence (see Collinge 2005). In this sense, it is claimed that notions of scale tend to hide the complex networks which stretch between purportedly local and global sites and connect and disrupt our established notions of scale (Murdoch 1997). Others claim that work on scale tends to reproduce a highly territorial vision of the world, which belies the aterritorial topologies (or connections) which constitute globalization (see Amin 2002). While I am sympathetic to many of these critiques, I do feel that they tend to unfairly represent the work of many scale theorists. While at one level the scale debate does find it hard to escape the seeming rigidity of the scales it describes, scale theorists have consistently emphasized the importance of understanding scales (like the neighbourhoods, regions or nations) not as absolute categories, but as relational entities. As relational entities, scale theorists argue that you can only begin to understand individual scales when you understand the social, economic and environmental relations which connect them. So for example it is only really possible to understand a regional scale by interpreting the relationship that a regional space economy has to its constituent metropolitan economies, and in turn the way the region then connects to national political arenas and directives (see Chapter 6). Additionally, while those working on scale have emphasized the abstract territorial boundaries which scales tend to construct, they have consistently argued that such scales are not eternal categories, but are flexible constructs, designed and configured as emergent ways of managing social, economic and environmental relations. The remainder of the book seeks to explore the sustainable society through its emergent scalar categories. Consequently, while globalization provides a crucial context within which I interpret the sustainability, it is not used to suggest that the sustainable society is a post-geographical world within which territorial boundaries (of all kinds) no longer matter. Instead through an exploration of the scales associated with sustainable development, this book charts the emergent geography of the sustainable society. As we will see, this emergent geography is sometimes based upon entirely new scales, and at others on the reinvention or reconfiguration of much older scales of social, economic and political life. In order to begin this scalar journey, however, the remainder of this chapter is dedicated to an exploration of the new global, or supra-national, scales which are becoming synonymous with sustainability. In particular analysis is concerned with how and why the ideal of the sustainable society is connected to visions of globalization and global political co-operation.

Globalization and sustainability

Before moving on to consider specific examples of how the processes of globalization have intersected with the rise of policies for a more sustainable society, I want to briefly consider why globalization and sustainable development are related. A number of writers have explored the relationship between globalization and sustainable development (see in particular Held *et al.* 1999: chapter 8; Sachs 1999). In order to begin to understand the connections between globalization and sustainability, it is necessary to unpack the notion of sustainability into its component environmental, social and economic parts. This section consequently focuses on the links between globalization and sustainability through a consideration of the globalizing dynamics of environmental change, social justice and economic development (for a broader discussion of these processes see Yearley 1996).

As we have discussed at the beginning of this chapter (and at some length in Chapter 1), environmental considerations have played a crucial role in our growing consciousness of the notion of globalization. This link between globalization and the environment can be understood in relation to three key considerations: (1) a growing awareness of the global processes which make the Earth an integrated environmental system; (2) the scientific production of new evidence showing that environmental change and degradation is increasingly moving from being local and regional phenomena to a truly global issue (to the extent that environmental transformation in one place is having impacts on other more distant environments and societies); and (3) debates regarding the use and management of global environmental commons.

With respect to the first consideration, it is clear that since the late 1960s and the Apollo space missions, there has been a renewed desire to understand the Earth as an integrated environmental system. This renewed desire has, however, not only been based upon the cultural impacts of pictures of the Earth from space, but also on new forms of imaginative science. While numerous scientists have dedicated their lives to studying the processes which connect environmental systems at a global level, perhaps the most famous exponent of this brand of global environmental science is James Lovelock. In his now well-documented *Gaia hypothesis*, Lovelock claimed that life on Earth is the product of a series of intricate balances and symbiotic relationships between the biosphere, atmosphere and hydrosphere. Crucially, Lovelock argued that these interdependencies could only be appreciated fully at a global environmental level – a level at which a homeostatic environmental balance has been achieved. One of the key implications of Lovelock's hypothesis, of course, is whether, if life on Earth is the product of a careful balance of environmental systems, human induced changes to the natural environment could threaten the ability of the planet to support life. While Lovelock's methods have been routinely maligned, his notion of a global environmental system – operating in a similar way perhaps to a human body – has been a great inspiration to many in the green movement.

In terms of the second point, the precise relationship between globalization and environmental degradation is difficult to assess. In his highly original book *Planet Dialectics* (1999) Wolfgang Sachs provides a timely analysis of the relationships between globalization and sustainability. According to Sachs, it is possible to link emerging forms of global environmental problems (like global warming, ozone depletion and marine pollution), which cross national boundaries, to key economic processes which also transcend national spaces (1999: chapter 8). At one level, Sachs argues

Box 5.1 James Lovelock and the Gaia hypothesis

James Lovelock developed his Gaia hypothesis while working on the Mars Viking Programme for NASA. In many ways his understanding of the reasons why planet Earth has an abundance of life was related to his claim that the planetary environment of Mars could not support any. His Gaia hypothesis was set out in his most famous book, *Gaia: A New Look at Life on Earth* (Lovelock 1979). By studying the operations of large ecological spheres, like the atmosphere, biosphere, cryosphere and hydrosphere, Lovelock charted the interconnecting balances which these systems had forged throughout environmental history. These balances, between global levels of oxygen and carbon dioxide, and marine salinity and cryospheric transformations, he argued, are the reasons why the Earth is able to support life. On the basis of these insights, Lovelock claimed that the Earth could be thought of as a giant super-organism, which as with other organisms had achieved a balanced, life-producing state. Lovelock's choice of the term Gaia is significant in this context, originating as it does from the Greek word for the goddess of the Earth.

Key reading: Lovelock, J. (1979) *Gaia: A New Look at Life on Earth*, in Lovelock, J. (1988) *Ages of Gaia*.

that the new, global mobility of economic activities (facilitated by telecommunications and new flexible production systems) means that an increasingly large expanse of the Earth's biosphere is now open to economic exploitation and resource extraction (Sachs 1999: 137). Beyond the physical exploitation of the environment, however, Sachs also notices how globalization has been synonymous with the spread of economic values developed predominantly within the economies of Europe, Japan and the USA. Consequently, with the deregulation of global trade and investment between nations, not only has more of the world's environment been opened up to economic use, but it has also been subjected to the *resource intensive*, fossil-based economic practices of the West (ibid.: 137–142). Based upon unfettered economic growth, competition and fossil-based growth, this is precisely the type of economic formula which many argue has caused contemporary global warming and its potentially hazardous socio-environmental consequences. Finally, Sachs argues that under contemporary globalization we have not only witnessed the spread of neo-liberal economic practices, but also certain cultural norms and values. In this context, throughout the world we see the same multinational corporations selling global goods (like motor cars, hamburgers and televisions), which are being mass consumed and incorporated into an increasingly wide range of people's lifestyles (Sachs 1999: 137). The spread of such global goods, and the types of lifestyles they offer (particularly with regard to the motor car), have the potential to deepen contemporary patterns of global environmental change. The point is that contemporary global environmental problems are not just part of the processes which are making us think about globalization, global ecological threats are actually a product of globalization and the economic practices and customs it is based upon.

The third and final intersection between the environment and globalization I want to consider relates to the management of global environmental commons. In many ways the growth of the global capitalist economy has been based upon the uncosted exploitation of environmental commons like the atmosphere and the oceans. By their very nature, these commons are environmental assets which do not, and cannot, belong to any one nation state. Because of the lack of a clear ownership system, however, global commons are often exploited by different nations and corporations to serve their own narrow economic and social needs. According to Garrett Hardin's infamous *tragedy of the commons* thesis, the unregulated exploitation of common resources could lead to the disintegration of the entire resource base, making key resources eventually unusable for all groups (Hardin 1968). While Hardin argued for the abandonment of common ownership systems in place of privately enclosed nature, the problems of commons management have actually resulted in a series of international efforts to co-ordinate the political and economic management of common resources throughout the world. Led by organizations such as the United Nations and World Wildlife Fund (hereafter WWF), the last thirty years has seen a proliferation of international conferences, protocols and treaties devoted to protecting environmental commons. Many interpret the taking of multilateral political responsibility for the management of the global commons in this way as an important political component of the myriad processes associated with contemporary globalization.

So we can see that in terms of its environmental concerns, the notion of sustainability is connected to the processes of globalization in a variety of important ways. In order to understand the nature of the relationship between globalization and sustainability, however, we must also appreciate certain key economic and social factors. As we discussed in the previous section, many interpret globalization as the historical extension of capitalist economic relations (see Harvey 2000). But as we saw in Chapter 4, the global spread of capitalist relationships throughout the world has been a highly unequal process. Consequently, while the operation of multinational corporations and trade organizations has opened up non-Western economies to new forms of investment, this investment has come at a cost. Whether it is in terms of the overt exploitations of the colonial era, or the more subtle modes of power associated with the *new world order* (see Hardt and Negri 2000: chapter 1), globalization appears to have strategically favoured Western market economies. The uneven development of capitalism can be seen today in the great social and economic disparities which exist between Western market economies and less economically developed countries. These inequalities in social and economic welfare are sustained by uneven trading relations, market protectionism, crippling debt and the saturation of LEDCs' economies with cheap surplus goods from the West. They are also manifest in elevated levels of social poverty throughout large parts of the non-Western world. The ideologies of sustainable development have become embroiled within the debates surrounding globalization precisely because notions of sustainability recognize that just as environmental problems cross national borders, many forms of social and economic inequality originate within global socio-economic relations. It is in this vein that advocates of sustainability argue that a sustainable society can only be achieved when the combined environmental, social and economic forces which are threatening life on this planet are recognized and addressed in a holistic way (see Chapter 4).

I started this chapter by emphasizing that globalization is not one thing, but actually a set of competing discourses. This sentiment is clearly echoed in Wolfgang's Sachs analysis of globalization and sustainability when he states:

> Globalization is not a monopoly of the neo-liberals: the most varied actors, with the most varied philosophies, are caught up in the transnationalization of social relations [. . .] Accordingly the image of the blue planet – that symbol of globalization – conveys more than just one message.
>
> (Sachs 1999: 153)

In this chapter I want to argue that sustainability represents a different message about what globalization can be. I want to claim that certain versions of sustainability offer a different vision of globalization than those promoted within free-market liberalism. This is a vision which is based upon social justice and ecological conservation, not unregulated economic expansion.

Climate change and the global response

In many ways the first example I want to consider of the relationship between sustainability and globalization is the most directly illustrative case of the complex social, economic and environmental forces which make up globalization. It is also, without question, one of the most widely discussed and debated issues within the policies and politics surrounding sustainable development. Climate change is a 'hot' political topic. It is the subject of numerous scientific studies, the object of an increasing number of TV documentaries, and has dominated the agenda of several major international conferences, from the United Nations World Summit on Sustainable Development (Johannesburg) to the G8 Summit of 2005 at Gleneagles, Scotland. Climate change is an important example of the relationships between sustainable development and globalization because at one and the same time the atmosphere is one of the most important global environmental systems – the recipient of many of the polluting outputs associated with economic development and a crucial basis for human life and health on this planet. As a meeting point (or perhaps context) for global environmental, economic and social processes and needs, it is hardly surprising that what goes on in the Earth's atmosphere has been a keenly contested area within contemporary discussions of the sustainable society.

The global atmosphere is a complex composite of gaseous elements, swirling winds, weather systems and convoluted pressure dynamics. While an important environmental system in and of itself, the global atmosphere cannot be understood as a closed system – it instead embodies a complex interface between the hydrosphere and biosphere and witnesses the continual exchange of elements and compounds between these systems everyday. The contemporary atmosphere is largely composed of nitrogen and oxygen, but it is also home to a range of important trace elements and compounds, including nitrous oxides, sulphur dioxide, ozone, methane, lead and of course carbon dioxide. The importance of the atmosphere to life on Earth is illustrated well in Bill Bryson's reflections on the planet:

> Thank goodness for the atmosphere. It keeps us warm. Without it, the Earth would be a lifeless ball of ice with an average temperature of minus 50 degrees Celsius. In addition the atmosphere absorbs or deflects incoming rays of cosmic radiation, charged particles, ultraviolet rays and the like. Altogether the gaseous padding of the atmosphere is like a 4.4-metre thickness of protective concrete and without it these visitors from space would slice through us like tiny daggers.
>
> (Bryson 2004: 313)

In addition to protecting us from dangerous forms of radiation, then, the atmosphere also ensures that the Earth is warm enough to sustain life. It is in this context that many people find global warming and climate change confusing.

Global warming is a naturally occurring phenomenon, driven by carbon dioxide in our atmosphere. In essence carbon dioxide acts like a giant blanket (or greenhouse) sealing in the Sun's heat, the heat which is reflected from the Earth's surface, and keeping the planet warm. To add to the confusion, not only is the greenhouse effect (at least partially) a natural phenomenon, but so too is climate change. Environmental scientists, glaciologists and climatologists constantly remind us of the great fluctuations which have occurred in the Earth's temperatures over time, as the planet has swung between ice ages and inter-glacial warmth. So you may well ask why all the fuss about contemporary patterns of climate change? Does it not simply represent the atmosphere doing what it is supposed to do? Should we simply accept climate change as a condition of life on Earth? Isn't trying to regulate the temperature of the planet an expensive and futile thing to do? Some people would answer yes to these last three questions (see Lomborg 2001). The problems with these assumptions start to emerge when you compare certain key atmospheric trends. Temperature records show that the twentieth century was on average a much warmer century than the nineteenth century; that the 1990s were the warmest decade so far on record; and that 1998 was the warmest year since records began.[2] Furthermore, the Intergovernmental Panel on Climate Change (IPCC) – the independent collection of scientists created to monitor global climate change – estimate that by 2100 the Earth will experience temperature increases of somewhere between 2 and 4.5 degrees Celsius (Lomborg 2001: 264).

The case for increasing global temperatures may appear conclusive, but you are entitled to ask is this just part of natural fluctuations in global temperatures, changes which have been occurring on this planet long before humans arrived? (For an interesting discussion of the social construction of climate change by a geographer see Demeritt 2001). We could take refuge in this question if it wasn't for the untiring work of one scientist in particular. Working from his observatory on the Hawaiian island of Mauna Loa, the climate scientist Charles David Keeling monitored levels of carbon dioxide in the Earth's atmosphere for many years. While carbon dioxide only makes up a tiny fraction of the atmosphere's gaseous content, it is one of the most important trace gases when it comes to global warming. Keeling started monitoring carbon dioxide in the late 1950s and his work showed a steady increase in the levels of the gas residing in the Earth's atmosphere (allowing for seasonal variations). Now at one level Keeling's finding were not unexpected; scientists recognized that large-scale industrialization had seen elevated levels of carbon dioxide being emitted into the atmosphere. But before Keeling's analysis, many felt that this excess carbon dioxide would simply be reabsorbed into the Earth's carbon cycle. Keeling's work indicated that not only was more carbon dioxide entering the atmosphere, but that it was also residing there over a long time period.[3] When Keeling's graph of steadily increasing atmospheric carbon dioxide is placed alongside data showing increasing global temperatures, the cause of global warming starts to look less like a natural event and more like a human-induced phenomena (see McKibben 2003). Debate still rages about the relationship between observed increases in atmospheric carbon dioxide and global temperatures. Could the two trajectories just be coincidental, or are they intimately connected? Those scientific experts working for the IPCC appear to believe the latter scenario, arguing that 'the balance of evidence suggests that there is a discernable human influence on global climate' (quoted in Lomborg 2001: 266).

Even if we accept (and many still do not) that global warming is a product of human environmental intervention, the consequences of global warming are still far from clear (see Figure 5.1). While climate change is a global phenomenon, one thing which is clear is that the projected effects of global warming will affect different geographical areas in very different ways, with often the most marginal of communities experiencing much more of the adverse consequences. The possible consequences of global warming include the loss of productive agricultural land, more frequent incidents of extreme weather events, sea-level changes and increased levels of inundation, and even the wider spread of tropical diseases (see United Nations 2003b; for a critical review see also Lomborg 2001: 287–297). Beyond the specific effects of climate change, however, it is crucial to realize that our contemporary social, economic and political systems are amazingly dependent upon climatic stability. Consequently, whether it be highly prized and valued real estate on low-lying ocean frontages, agricultural communities in Bangladesh, or even the electrical cables which power our homes (there is a fear that heat expansion caused by global warming could seriously damage such supply lines), much of the sustainability of our current ways of life depend on the climate not changing.

In response to these perceived threats, the last twenty-five years have seen a range of international initiatives emerging which have sought to develop more sustainable socio-economic relationships with the global atmosphere. At the centre of these initiatives has been a desire to try and instigate a global response to these global environmental problems. A global level response to climate change is seen by many as vital, because even if certain countries reduced their production of carbon dioxide, greenhouse gases from other countries could continue to drive climate change and effect other people's and nation's livelihoods throughout the world. The first major international agreement on climate change abatement was formulated at the Rio Earth Summit in 1992. The United Nations Convention on Climate Change took effect on 21 March 1994, after it had gained the signature of 165 national leaders. While this convention has still

Figure 5.1 The possible consequences of global warming (taken at the Centre for Alternative Technology, Wales)

not been ratified by all of the countries that signed up to the convention, its main aim was to stabilize global carbon dioxide production so as to enable global ecosystems to adjust sustainably to climate change.[4]

In 1997 the Kyoto protocol superseded the United Nations Convention on Climate Change. As its name suggests, the protocol was developed in the Japanese city of Kyoto. Its basic aim was to reduce the amount of greenhouse gases produced by signatory states by an average of 5.2 per cent below their respective 1990 levels. This target has to be achieved during the time period running from 2008 to 2112, or else failing states are subject to further greenhouse gas reduction penalties.[5] Although the fact that the Kyoto protocol was signed by over one hundred states, its effectiveness as a form of international collaboration to tackle climate change has been questioned for some time. In order to come into force, the Kyoto protocol needed to be ratified by states that collectively produce 55 per cent of global greenhouse gases. For a long time, the reluctance of nations like Australia, the USA and Russia to ratify the agreement meant that the protocol was not binding on the other states that had signed it. In October 2004, however, the Russian government finally ratified the Kyoto protocol and reignited the long-stalled process of international action on climate change. The protocol has still not been ratified in the USA. Opposition to the protocol in the USA has been expressed in a number of ways. First, the George W. Bush administration continues to question the scientific link established between rising levels of industrial greenhouse gases and climate change. Second, the US government is also concerned about the economic impact which ratifying the Kyoto agreement could have on the country (Lomborg for example estimates that the Kyoto protocol could cost the global economy approximately US$150 billion a year, 2001: 322) (see Chapter 2). The USA remains the world's largest emitter of greenhouse gases. While this remains the case, and the USA continues to dispute the principles of the Kyoto protocol, it appears that global attempts to tackle climate change will remain seriously compromised.

Interestingly, in the context of this book, the Kyoto protocol incorporates within its policy dictates an appreciation of the variable geographies of sustainable development. Recognizing the urgent need for economic development in many LEDCs, the Kyoto protocol only sets strict targets for carbon dioxide emissions reductions in MEDCs, while offering more flexible mechanisms for carbon reduction for other nations. The geographical sensitivities associated with the Kyoto process have, however, created further geopolitical tension, with the USA arguing that large carbon dioxide emitters such as China and India need to make firmer commitments to reductions before they can themselves consider commiting to the protocol. The scientific, political and economic issues surrounding climate change are complex and it is not the purpose of this section to put forward an unqualified assault on US climate change policies (it is important to realize for example that despite US opposition to the Kyoto protocol at a federal level, many individual states have begun legislating to reduce greenhouse gas production at a state level – most famously perhaps in Governor Arnold Schwarzenegger's state of California, see Toepfer 2004). What these discussions reveal, however, are the geopolitical difficulties associated with developing global action on sustainable development issues.

The story of climate change reveals many of the tensions which exist between the processes of globalization and sustainability. At one level it appears that the economic processes associated with globalization are generating contemporary patterns of climate change and threatening global socio-environmental sustainability. At the same time, it appears that effective forms of global political action are required to tackle the socio-environmental threats associated with global warming. Such action it is argued could impinge upon economic growth in both MEDCs and LEDCs, and threaten socio-economic sustainability worldwide. It is in this context, that despite being related to global phenomena, the principles of sustainability, and debate concerning how to achieve the most sustainable solution to global problems, have their own geographies. Sometimes this geography is expressed at a national level, with countries like the USA refusing to ratify the Kyoto protocol. At other times, the geographical response to sustainable development issues takes a more local form, as individual neighbourhoods, cities and regions decide to address climate change in their own ways and through their own programmes (see Burgess 2005; Slocum 2004). The point is that globalization does not represent the end of geographical enquiry into sustainable development issues, but rather the need for a newly intensified study into the emerging spaces and scales of sustainability (see Eden 2005).

Supra-national co-ordination and the case of the Barents Sea region

Emerging socio-environmental conflicts in the Barents Sea

As we discussed in the first section of this chapter, it is important not to think of globalization as only involving global-level events. Globalization should instead draw us in to considering the reconfigured scales and spatialities associated with contemporary socio-ecological life and existence. The remainder of this chapter considers one example of sustainable development which, while not operating at a 'global' level, is clearly occupying a space which has been opened up by the complex processes associated with globalization. The Barents Sea is a remote expanse of ocean located to the north of the Arctic Circle and which meets the shores of Norway and Russia (see Figure 5.2). For a long period of time the relative remoteness and inaccessibility of the Barents Sea has meant that it has remained a relatively unspoilt ecological region (for a discussion of the geopolitics of the Barents Sea see Churchill and Robin 1992; Osherenko and Young 1989). Recent developments have, however, changed popular understandings of the Barents Sea and seen it take an increasingly prominent role in debates surrounding globalization and sustainability. Two key events have been central to this process. First, scientific surveys have indicated that beneath the Barents Sea lies the world's largest untapped oil and gas reserves. As two of the main energy sources driving the global economy, excavation of these reserves could be extremely beneficial to the economies of Norway and Russia. Second, Russia has started to explore the potential of building an oil pipeline to Murmansk (the main Barents Sea port), through which oil extracted from continental Russia could be transported via oil tankers all around the world. These twin developments have generated a great deal of anxiety among many environmental groups. Organizations such as the WWF have described the Barents Sea as one of the world's few unspoilt fragments of marine nature and argue that the extraction and transport of oil through the sea could jeopardize the delicately balanced ecosystem of the area (see Figure 5.3).

Figure 5.2 The Barents Sea

Figure 5.3 Konigfjord, Svalbard

In the context of the struggles surrounding the use and management of the Barents Sea, I want to argue that the region represents an important meeting place of globalization (in the form of trade, resource extraction and oil transportation) and sustainability (expressed in terms of the environmental integrity of the area). The story I am going to tell about the Barents Sea, however, will reveal how in the context of the late arrival of globalization in the region, the Barents Sea has been re-forged as a supra-national zone of resource conflict and environmental management. In this way we will see that various attempts to claim the Barents Sea have involved the reconstruction and re-imagination of its geographical parameters.

Sustainable development planning in the Barents Sea and the WWF

In many ways the tensions over resource development in the Barents Sea illustrate the geographical nature of sustainable development issues. In one geographical region, the Barents Sea, we find the presence of untold riches of fuel and of ecological diversity. To extract oil and gas could be vital for economic development and social employment in both Russia and Norway. But to do so could also threaten the unique marine ecosystem and environments which surround the oil reserves, through the building of industrial pipelines and the dangers of large-scale oil spills. Tensions surrounding economic and environmental sustainability often arise as a result of such geographical conundrums. In response to concerns over sustainable development in the area, the WWF has been trying to think of alternative ways in which the Barents Sea can be thought about and used. Significantly in the context of this book, the WWF have developed a sustainable management strategy for addressing resource conflict and ecological conservation in the sea.

According to the WWF, environmental management in the Barents Sea has been inhibited by a very narrow conception of what the Barents Sea actually is. At one level this narrow conceptualization of the Barents Sea has been based upon understanding it primarily as the home of energy resources. Despite the continuing political obsession with fossil-based resources in the region, the WWF have worked hard to uncover an amazing wealth of ecological resources and diversity in the area. A biodiversity assessment of the Barents Sea (conducted by the WWF) revealed the wealth of biodiversity in the area (see WWF 2001). According to the WWF's report, the Barents Sea represents a highly *productive and fluctuating* environment. With its unique mix of ocean currents, ice sheets and nutrient flow, the Barents Sea is able to play host to a range of interconnected organisms and species (ibid.: 22–24). In addition to a range of fish life, this biodiversity includes ice flora and fauna, plankton, whales, walruses, seals and various varieties of sea bird (ibid.: 25–48). In addition to this rich biodiversity, the Barents Sea region is also home to two groups of indigenous peoples, who are rarely mentioned in the prevailing economic discourses of the area. The region is home to several tens of thousands of Sami and Nenets (ibid.: 25). By highlighting and recording this environmental and social data, the WWF are trying to promote a vision of the Barents Sea as a space of social, ecological and economic interest. The second reason why those working for WWF believe that the Barents Sea has been narrowly misconceived is because of its territorial division. According to the WWF, the reason why such ecological diversity exists in the Barents Sea region is because of the forms of environmental and biological interaction which occur throughout the whole region. They further claim, however, that the division of the region between capitalism and communism during the Cold War and now between the resource interests of Norway and Russia, is making it hard to think and act on the region as an integrated socio-environmental whole (WWF 2004).

In the context of this narrow and territorially divided vision of the Barents Sea, the WWF proposed a new spatial management strategy for the region (see WWF 2004). This strategy, known as *the Barents Sea Ecoregion Programme*, was launched in 2004. It is important at this point to explain what is meant by the term ecoregion as it is being deployed by the WWF. As we will see in Chapter 6, as a spatial entity, regions usually refer to sub-national territorial entities, which are used to divide up larger states into political, cultural, economic and environmental districts. This is not, however, how the WWF are currently using the term. In their Barents Sea campaign, the WWF are utilizing the idea of an eco'region' in order to construct a supra-national space, which cuts across Norwegian and Russian territorial sovereignty in the area. In this way the ecoregion is trying to create a sense of environmental space which existed in the Barents Sea long before the emergence of modern nation states. According to the WWF, the Barents Sea ecoregion incorporates both the Norwegian and Russian sections of the Barents Sea as well as large parts of the terrestrial hinterland which contribute to the ecological balance of the area (WWF 2004).

As an international non-government organization (a type of organization which appears to be having an increasingly important role within the era of globalization, see Desforges 2004), the WWF are of course not able to enforce their policies and visions for the Barents Sea. In order for the ecoregion strategy to be enacted the WWF require the full co-operation of the Norwegian and Russian authorities. At present, while the WWF have gained support from the Norwegian government, negotiations are still ongoing with the Russian government. If delivered, however, the WWF's vision of ecoregional development would see the Barents Sea managed as an integrated social, economic and environmental space. According to this plan, any form of proposed economic development in the region would have to be carefully assessed for its potential social and ecological consequences. Research on socio-environmental relations in the Barents Sea indicates that the implementation of new modes of integrated sustainable development cannot come soon enough (see Brunstad *et al.* 2004). With extensive industrial development already occurring in the Russian Barents and plans to extend oil supply and refining facilities there, the future sustainability of the area is already under serious threat.

While obviously not representing a global response to socio-environmental problems, the WWF's ecoregion is typical of the types of supra-national spaces of sustainability that are emerging in the shadow of globalization. Such supra-national spaces of sustainable development planning can encompass environmental units like rainforests, or economic units such as the European Union. What these spaces all have in common, however, is a desire to implement sustainability in ways which cut across established territorial boundaries, in order to produce more integrated spatial systems of planning and development.

Summary

In this chapter we have explored the multiple meanings associated with the term globalization and how it is intimately connected to our discussions of sustainability. At its simplest level we have seen that the multiple processes associated with globalization force us to think beyond the narrow bounds of nation states in our attempt to understand how the world works. In relation to sustainability, we then moved on to consider how the social, economic and environmental processes which determine our relative levels of sustainability are in part constituted at a global level. From the systemic forms of environmental change associated with climate change, to the imperialistic economic relations of trade which cause so much social inequality in the world, we have seen the inextricable links between globalization and questions of sustainability.

Despite our explorations of globalization, this chapter has not suggested that we should see globalization as a form of weightless, aterritorial existence which signals the end of geography as we know it. Instead, this chapter has revealed that the processes of globalization are creating a series of new geographical spaces and scales (see also Chapter 1). While many of these new spaces and scales are devoted to narrow economic goals, as our two case studies have shown, certain spaces are emerging in this era of globalization which are being designed specifically to deliver a more sustainable future for our planet. In our first case study, we saw how contemporary attempts to tackle the serious social, economic and environmental threats of globalization are operating through global networks of international agreements and protocols. In this sense, the global atmosphere is becoming a meeting point for different components of the sustainable development debate, within which the needs of oil-based industries are colliding with those concerned about the social utility and environmental integrity of the atmosphere. In the second case study, we saw how the pressures associated with global economic development are increasingly threatening biodiversity and delicate ecological systems throughout the world. In the case of the Barents Sea, we saw how such tensions are in part being fuelled and sustained by narrow nationalist claims to territorial resources. The activities of the WWF, however, serve to illustrate how certain NGOs and political groups are using the trans-national visions, typically associated with globalization, to construct new supra-national spaces for sustainable development. What is perhaps most interesting about the activities of the WWF in the Barents Sea is that it is still dependent upon the active participation of territorial powers – as is still the case with climate change negotiations. In this context, it appears unwise to simply acclaim the arrival of a more integrated global space within which sustainable development can thrive, and to remember to be vigilant about the sustainable development credentials of the new spaces which are emerging in the contemporary global era.

Suggested reading

There is a bewilderingly wide range of introductory texts which can be used to familiarize the reader with the key concepts associated with the term globalization. There are, however, fewer texts which effectively link globalization with debates surrounding sustainability. Held, D., McGrew, A., Goldblatt, D. and Perraton, J. (1999) *Global Transformations – Politics, Economics and Culture* provides a good overview of a variety of themes relating to globalization and also includes one chapter on the links between globalization and environmental issues. Sachs, W. (1999) *Planet Dialectics: Explorations in Environment and Development*, chapter 8, provides one of the best critical discussions available on the links between globalization and sustainable development.

Suggested websites

For an accessible discussion of the evidence, policies and possible consequences of global warming, visit the BBC's climate change page at: http://www.bbc.co.uk/climate/

For more information on key international policies designed to tackle climate change go to the United Nations Environment Programme's climate change page at: http://www.unep.org/themes/climatechange/

Information on the WWF's Barents Sea campaign is available at: http://www.panda.org/about_wwf/where_we_work/arctic/what_we_do/marine/barents/index.cfm

6 THE SUSTAINABLE REGION

Introduction: a tower with a regional view

Towards the end of the nineteenth century, the sociologist and planning theorist Patrick Geddes opened the Outlook Tower on Castle Hill in Edinburgh. The tower, which had previously been the town mansion of the Laird of Cockpen, and more recently an astronomical observatory, was five storeys high and had a small viewing terrace and camera obscura. Geddes divided his tower into different rooms, one dealing with the solar system and the cosmos, one with the planet Earth, another with Europe, one devoted to Scotland and the other to the city of Edinburgh itself. The purpose of the tower was to provide the citizens of Edinburgh, and more distant visitors, with a geographical device for understanding Edinburgh's place in the world. In his guide for visitors who wished to explore the tower, Geddes suggested that they begin at the top and work their way down through the subsequent rooms (Geddes 1906); Geddes wanted visitors to begin their explorations of the tower on its roof terrace. This starting point was important to Geddes, because the roof terrace enabled people to escape the built-up congestion of the street and to see the city of Edinburgh within its wider 'regional context' (as we see in the next chapter, this is not physically possible with many larger cities). To Geddes, gaining a regional perspective was important because it forced urban citizens to think beyond the bounds of their urban communities and to recognize the broader social, economic and environmental processes which supported a city. On the roof terrace, Geddes believed that the visitor would be better able to comprehend the geological, agricultural and socio-economic hinterlands which supported the city. According to Geddes, on the roof terrace the visitor was forced to be a meteorologist, geologist, botanist and geographer. But what is particularly interesting about Geddes's Outlook Tower from the perspective of this book is that it was designed to endow people with a regional consciousness which was founded upon a keener awareness of the geographies of socio-environmental interaction (interestingly, adjacent to the roof terrace, Geddes had set aside a small room – designed for a single occupant – where the visitor could sit and silently reflect upon their place within regional space).

I start with this discussion of Patrick Geddes's Outlook Tower partly because of my own fascination with the figure of Geddes himself but also, and perhaps more importantly, because this story eloquently reveals many of the reasons why the region has become such an important spatial category within contemporary debates concerning the sustainable society. The first thing we can derive from this story of the Outlook Tower is what we mean by the term region. The fact that Geddes believed that people required the elevated perspective of the tower to see a region illustrates that regions are typically understood as spatial scales which are larger than cities, but smaller than nation states. As *meso*, or middle, scales which occupy the spaces between nations and localities, regions can actually take many different forms (Soja 2004). Regions can be political or administrative (Jones and MacLeod 2004); have an economic basis (Storper 1997); be linked to issues of sub-national cultural identity (Paasi 1991); or be identified in terms of certain environmental features (Pepper 1996: 306–309). The different forms which regions can take make them confusing and even infuriating categories for students to study. The confusion surrounding the notion of the region is further compounded by the fact that despite Geddes's notion of a regional view (from atop of his Outlook Tower), regions are actually only very rarely things that we can actually see in the real world, with definite and visible boundaries. The second point to notice, however, about Geddes's vision of the region, is that it represents a space or scale where it appears to become possible to comprehend the ways in which social practices, economic development and environmental processes are connected. As we will see, the idea of the region as a space of socio-ecological interdependency has been a crucial factor in the growing significance which has been given to the region in debates concerning the sustainable society.

Leaving Patrick Geddes and his Outlook Tower, the rest of this chapter is going to explore the connections between sustainability and regions. In this context, the chapter begins by considering how different geographers have used and interpreted regions and how the notion of a sustainable region both supports and challenges these understandings. The remainder of the chapter then considers two very different examples of sustainable regional development in practice. The first of these case studies considers the emergence of the Ecodyfi sustainable region in Mid Wales; the second explores sustainable regional development policies within the political region of the West Midlands (in central England). These two case studies illustrate the different meanings that are often attached to sustainable regions and the political processes which inform their construction.

Regional geographies and the political and ecological uses of regional space

Regional thought waves and the geographical tradition

For a long time I think I was oblivious to the existence of regions. In fact I do not believe that it was until I started to study geography at university that I began to realize that regions existed. I was born and brought up in the small town of Walsall, but it wasn't until I left that I realized this town was actually in a region called the West Midlands. When I first moved to Wales, I did not realize that Aberystwyth was actually part of a larger region. As part of a research project I was carrying out I later discovered that Aberystwyth was actually in two regional spaces – the Mid Wales region and a European Objective 1 region. All of a sudden regions started to appear everywhere. Some of my friends at university informed me that they came from the North East region of England; on visiting other friends who lived in Germany, I discovered that their home was in a region called Baden-Württemberg. My growing sense of regional space was, however, confused, because while I was suddenly aware of people talking about regions, I also realized that different people seemed to use the term region in very different ways. Sometimes my lecturers used the term region to refer to different parts of a city (like front and back regions); when John Major (the then British Prime Minister) talked about regions he seemed to be referring to large sub-national spaces like the south east of England or Wales; but when Kofi Annan used the term region, he did so to discuss much larger places like the Middle East or Central America. I was thoroughly confused!

While studying geography at university, I was able to come to terms more effectively with precisely what regions are. Initially I discovered that regions were one of the main objects of geographical enquiry and investigation. Indeed in the late nineteenth and early twentieth centuries, a group of 'regional geographers' devoted their time to uncovering regions through careful analyses of different landscapes and cultures (see Agnew *et al.* 1996). Famous regional writers like Vidal de la Blache analysed regions (or *pays*) like Alsace-Lorraine, and built up complex and highly sophisticated accounts of their varied environmental, social, political and cultural characteristics. Within this approach to regions, two assumptions appeared to prevail. First, there was a belief that regions existed, almost as natural entities and simply awaited discovery by the geography. Second, there was a strong assumption that regions were unique spaces, with definite boundaries which could be identified on the basis of their internal social, cultural and environmental unity. After the Second World War, and with the onset of the quantitative revolution in European and North American geography, regions were understood less as naturally occurring geographical entities and more as models which could be used to give statistical order to geographical space. This was the age of the standard, or statistical region, identified not because of its unique culture, but because of its similar demographic or economic dimensions to other regions (for more on the study of regions in geography see the collection of regional writings brought together in Agnew *et al.* 1996: 365–413).

What both the natural and standard visions of the region have in common is that they interpret regions as bounded, sub-national spaces. During the 1990s a new geographical approach to regional study started to emerge. Customarily referred to as the *New Regional Geography* (see Jones and MacLeod 1999; MacLeod 1998), this school of regional geography differed in two key ways from its predecessors. First, it set out to understand regions not as isolated internally integrated spaces, but to explore how regions are made through their social, economic and political relations with other places. Second, this new approach to regions sought to emphasize that regions did not exist as natural entities, but were socially constructed spatial categories (see Allen *et al.* 1998). During their innovative analysis of the South East region of England, for example, Allen *et al.* assert that, '[r]egions only exist in relation to particular criteria. They are not "out there" waiting to be discovered; they are our (and other's) constructions' (1998: 2). The idea promoted within the New Regional Geography that regions are socially and politically constructed, is not meant to deny that regions are being forged and demarcated all of the time. Rather this perspective on regions is meant to make us question how and why regions are constructed, and how those constructions can vary in both time and space (see Box 6.1). Again in this context Allen *et al.* remind us that '"regions" (more generally "places") only take shape in particular contexts and from specific perspectives. There will always be multiple, coexisting characterizations of particular spaces/places' (1998: 34).

The sensitivity of the New Regional Geography to the construction of regions has important implications for our discussions of the sustainable region. As we will see in the next section, regions are crucial spatial arenas within which contemporary attempts to build more sustainable societies are being waged and contested. But it is important to ask how and why this sub-national spatial scale is being used within contemporary policies for sustainable development. In this context, this chapter is really interested in asking what affect naming and constructing regions as sustainable is having on regional forms of socio-economic development and environmental management (see Chapter 5 on the eco-management of the Barents Sea).

Box 6.1 The discontinuous region: on the English South East

In their analysis of the South East region of England, Allen *et al.* (1998) challenge traditional visions of regions as natural units. Over the last twenty-five years, the South East region has been the fastest growing economic region in the UK and has been the subject of much discussion regarding regional planning and identity. In their book, *Rethinking the Region*, Allen *et al.* question the basis upon which a region called the South East actually exists. Despite the fact that a politically designated standard South East region has been demarcated by the British state, Allen *et al.* expose the contradictions and complexities that this designation contains. They question, for example, whether any region that includes the English Home Counties (that bastion of *Middle England*, or the quintessential heart of the English nation) and the London inner-city boroughs (with their mix of working-class and ethnic communities) can be considered an integrated regional space. They also reveal how the idea of where the South East as a region actually is has varied greatly over time. Essentially, despite writing a book on the South East region, Allen *et al.* find it impossible to settle on where that region actually begins and ends. In this context, the South East region is perhaps best thought of as a large geographical space which very few of the people actually living within this area identify with. The region has thus been established for largely political not cultural or environmental reasons (the region is essentially being used to regulate urban expansion and integrate economic policies). The political role of the region has, however, also been brought into question with the establishment of the Greater London Authority – an authority which seeks to tackle key policy issues in the south east of England through a set of urban not regional policies.

In this broad context, Allen *et al.* call for a more flexible view of regions as discontinuous (or internally fragmented) and unbounded (or open) places. In conceiving of regions in this way, Allen *et al.* reveal two important things: (1) that regions are not simply made from within their own territories, but are forged through relations of trade and cultural exchange with other places and regions; and (2) that the existence of any region is the product of a set of political struggles and contestations.

Key reading: Allen, J., Massey, D. and Cochrane, A. (with Charlesworth, J., Court, G., Henry, N. and Sarre, P. (1998) *Rethinking the Region*.

The beauty of the regional scale

Hopefully you now have a better idea of what regions actually are – at least within geographical study. Before I move on to consider the emerging significance of regions within visions of the sustainable society, I want to explore why the region is seen as such an important vehicle for sustainable development. In order to understand the importance of the region within contemporary debates over the sustainable society, we need to reflect upon the long-held suspicion of environmentalists about the ability of nation states to effectively manage the environment. In Chapter 1 we discussed the ideas of the famous green thinker Ernst Schumacher and his book *Small is Beautiful* (1973). In many ways Schumacher's arguments are part of a long tradition of green writings on the inadequacies of large-scale modes of economic and political organization to manage the environment. While Schumacher used his book to critically analyse large-scale economic systems, the nation state has also been widely criticized within environmental philosophy. On many counts it is anarchist thinkers who have developed the most concerted critique of the environmental effects of big government (although the writings of eco-socialists and neo-Marxists should not be discounted here). Famous anarchist writers like Michael Bakunin, William Godwin, Emma Goldman, Peter Kropotkin, Pierre-Joseph Proudhon, Elisée Reclus and Henry David Thoreau have all described the damaging social and ecological effects associated with the political hierarchies of national governments. Anarchists argue that the political hierarchies associated with nation states result in key state officials and departments taking responsibility for key dimensions of environmental management. The results of this state-based system of environmental management are twofold: first, important forms of local ecological knowledge are routinely ignored within large-scale environmental policies (see Scott 1998); and second, individual people tend to become alienated from the environment, as they see it predominantly as an area of state not individual responsibility.

In light of their inherent suspicion of the institutions of the state, anarchists have consistently called for the creation of smaller-scaled political communities within which individuals can feel more empowered and responsible for their own environment and better understand its role in supporting local economic development (see Kropotkin 1912). One of the most influential works on the problems of national governments was Leopold Kohr's *The Breakdown of Nations* (1957). Leopold Kohr was an Austrian philosopher who, amongst others things, was Ernst Schumacher's intellectual mentor. In *The Breakdown of Nations*, Kohr argued that giant nation states are ineffective at instigating the participation of civil society within political policies and inflexible in responding to changes in the economy and environment. Kohr consequently argued that smaller states (like Switzerland) and regions (like Wales) provided much more effective scales through which to deliver social, economic and environmental policies (see Palmer 2001: 209–210). It is in the light of the uncertainty which surrounds the ability of nation states to develop sustainable social and ecological futures, that the region has gained so much recent attention within discussions of the sustainable society. But what is it about the regional scale that makes it such an attractive scale around which to conceive of a more sustainable society? The remainder of this chapter explores this question by considering two examples of sustainable regional planning in practice. As both case studies illustrate, it is the nature of the region as a middle (or *meso*) scale which lies at the heart of its utility within discussion of the sustainable society (Soja 2004). As a meso scale, it is argued that regional forms of socio-economic organization are much more sensitive to local social and environmental needs than national governments. As scales which are, however, largely constituted above local communities, regions are also able to transcend the purported tyranny of localized, and often selfish, interests, to look to the long-term needs of both society and the environment.

The rural region and the Ecodyfi initiative (Wales)

Redefining the region

The first example of sustainable regional development I want to consider lies only eighteen miles from where I am currently sitting writing this chapter. In the heart of rural West Wales there has been an ongoing campaign to create a sustainable region (for more on the emergence of sustainability in rural areas see Marsden *et al.* 2001). This vision of a sustainable region was first formulated by an organization called Ecodyfi. Ecodyfi is a community-led regeneration scheme which operates in the Dyfi River Valley. The Ecodyfi partnership was developed during 1997 and finally established in 1998. According to its development plan, the aims of Ecodyfi are 'to foster sustainable community regeneration in the Dyfi valley [and] to deliver environmental, economic and social benefits simultaneously [while taking] a long-term and global perspective in meeting local needs' (Ecodyfi 2002: 1). What is unusual – and I would argue most interesting – about what is happening in the Dyfi Valley is that it represents an attempt to construct a sustainable region by the people actually living in the area, and that the region these people are constructing looks very different from what we might expect a sustainable region to actually be.

So what precisely is the Ecodyfi region? While defined in slightly different ways at different times, the region is now taken to be the twelve town and community council areas which cover the west of Powys, the southern tip of Gwynedd and the northern extreme of Ceredigion (Ecodyfi 2002) (see Figure 6.1). According to the Ecodyfi partnership (2002), this is an area of 600 km² which contains some 4,600 households and 11,350 residents (these figures are based on the 1991 census). What is most interesting about this designation of regional space is that it is a region which actually cuts across and largely ignores existing political borders. While the region is based upon council districts, the Dyfi Valley region actually transects four planning authorities – the county councils of Ceredigion, Gwynedd, Powys and the Snowdonia National Park Authority. In this context it is a region which has not been created for the convenience of existing planning authorities, but because it marks out a space which has social and environmental meaning to many who live there. One of the key factors underlying the designation of the Dyfi Valley region is that it covers the headwaters and main tributaries of the Dyfi River system (see Figure 6.2).

The use of regional space to mirror and manage environmental systems like river valleys, mountain ranges, or forest ecosystems is characteristic of something called *bioregionalism* (see Box 6.2). Bioregional philosophy suggests that regional boundaries should be constructed in ways which best

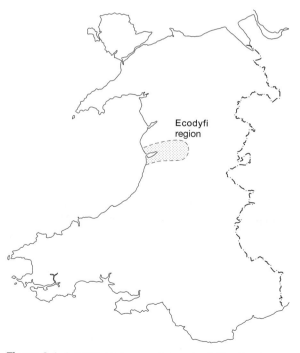

Figure 6.1 A socio-environmental region? The Ecodyfi region (drawn by Anthony Smith)

Figure 6.2 The ecological heart of the region: the Dyfi River Valley

serve the management and care of the environment. As a middle scale, regions lend themselves to these bioregionalist goals because they are small enough (unlike states) to be sensitive to environmental systems, but also large enough (unlike local authority districts) to cover whole ecological systems. So there is an obvious link between the vision of a sustainable region and the principles of bioregionalism. At a simple level this link can be discerned in the fact that as with sustainable regions, bioregions are designed to be socially and environmentally sustainable places. This sustainability is based upon a good regional understanding of local socio-environmental needs and a balanced approach to economic development. There is, however, an important difference between sustainable regions as they are currently being constructed (like that in the Dyfi Valley) and the vision of bioregions. This difference is based upon the fact that while sustainable regions are supposed to minimize their socio-ecological footprints in their relations with other regions, bioregions are supposed to form more autonomous regions, which are quite literally self-sustaining (see Box 6.2). Kirkpatrick Sale for example argues that bioregions should be able to provide for all their own resource needs and deal with all of their own pollution. This vision of bioregionalism is very different to the outward-looking and inclusive regional visions associated with sustainable regionalism.

Box 6.2 Principles of bioregionalism

While often criticized – and in many instances rightly – for its apparent whimsical nostalgia for a lost form of harmonious socio-ecological existence, it is clear that the principles of bioregionalism have had a significant affect on the idea of sustainable regional development. Bioregionalism is an intellectual and political movement which argues that the key to creating a more sustainable way of life is the restructuring of human settlement patterns, modes of political organization and economic practices in regional communities. The ideas of bioregionalism are most directly associated with the writings of Kirkpatrick Sale. Kirkpatrick Sale argues that the principles of bioregionalisms can be discerned from the word itself. Hence it is about *bio*logical issues (from the Greek word for life), regions (from the Latin for territorial space) and it is an *ism* (from the Greek for doctrine) (Sale 1980). In this context, bioregionalism can very simply be understood as a doctrine which suggests that the needs of human and non-human life are most effectively met through small-scaled regional communities of living. Bioregionalists consequently argue that in order for people to understand the place in which they dwell and the environment on which they depend, modes of social organization need to be scaled down. To quote from a lecture which Kirkpatrick Sale gave: 'We must somehow live as close to it [the land] as possible, be in touch with its particular soils, its waters, its winds; we must learn its ways, its capacities, its limits; we must make its rhythms our patterns, its laws our guide, its fruits our bounty' (1974). Bioregionalism is consequently based on a system of political decentralization, through which individuals are given greater say and control over their surrounding environment. It is also predicated on the construction of regions on the basis of environmental, not political and economic criteria. Finally, according to Kirkpatrick Sale, bioregionalism is about creating autonomous regional communities, which in their own unique ways learn to live and work with their environments in a sustainable way. An extreme version of bioregionalism would actually see regions which survive only on the resources which the region itself could provide.

Key reading: Pepper, D. (1996) *Modern Environmentalism: An Introduction*: 306–309; Sale, K. (1980) *Human Scale*; Sale, K. (1974) 'Mother of all', in S. Kumar (ed.) *The Schumacher Lectures Vol. 2*.

The unofficial sustainable development capital of Britain: reflections on the Ecodyfi initiative

In order to understand the ecologically inspired vision of sustainable regional development evident in the work of Ecodyfi, I want to conclude this section by looking more carefully at the history of the region and the contemporary initiatives which are in operation there. The seeds of sustainable regional development in the Dyfi Valley were actually sown long before the Ecodyfi partnership was even conceived of. The Dyfi Valley has a long historical association with sustainable development initiatives. In 1974, for example, the valley became home to the Centre for Alternative Technology (or CAT). This centre was set up by a group of environmental volunteers and was designed to promote sustainable lifestyles and new green technologies (see Figure 6.3). Then in the 1990s the Welsh Development Agency supported the foundation of the Dyfi Eco Park (see Figure 6.4). The Dyfi Eco Park was designed to provide a business park for green businesses. Built using sustainable architectural techniques, the Dyfi Eco Park remains an important centre for green business innovation in Wales (see Figure 6.5). This collection of sustainable development initiatives led to the Dyfi Valley being dubbed *the unofficial British capital of sustainable development*. In many ways the Ecodyfi partnership represents an attempt to draw these different local initiatives together in order to foster

Figure 6.3 Alternative lifestyles at the Centre for Alternative Technology (Wales)

Figure 6.4 and Figure 6.5 Green business and sustainable architecture in the Dyfi Eco Park

a broader regional movement for sustainable development in the valley. As we have already stated, the Ecodyfi initiative started in 1997. Ecodyfi actually comprises representatives from the local community, private businesses in the region, voluntary associations, different local authorities who have jurisdiction in the region, the Snowdonia National Park Authority and the Welsh Development Agency. From the start, the principle behind Ecodyfi was that it would provide a form of regional umbrella within which different local sustainable development initiatives could be supported and harmonized (Ecodyfi 2002: 2). Perhaps the key driving force behind Ecodyfi was a belief that life in the Dyfi Valley had become unsustainable. What is most interesting about Ecodyfi's understanding of the unsustainability of the area is that it was based not only on economic and environmental observations, but also on cultural issues. The Dyfi Valley is a strong Welsh-speaking and cultural community, and members of Ecodyfi were concerned that changes in the rural economy of the region were threatening its cultural sustainability. The decline of the agricultural sector in the area during the late 1980s and early 1990s, combined with poor-quality housing and a lack of alternative employment, led to the spread of rural poverty in the region. In turn this led to a situation whereby young Welsh-speaking members of the community were leaving the area, only to be replaced by older non-Welsh people migrating into the region (see Ecodyfi 2002: 5–7). In this socio-economic and demographic context, Ecodyfi felt that it was vital to regenerate the economic fortunes of the area if Welsh cultural life and language use were to survive. (As will become apparent in the next chapter on urban sustainability, the sustainable development priorities of rural areas are often very different to those of cities, being based much more on concerns with the threats of under rather than over development, and on the preservation of rural lifestyles and cultures.) The problem then became how to reconcile the need for economic development with the protection of the unique environmental assets of the region.

From the very beginning of the Ecodyfi partnership there has been a strong emphasis on the need to balance social, cultural, economic and environmental considerations, and thus not to be seen simply as an environmental protection agency. Ecodyfi have been keen to highlight that the 'Eco' prefix of their name is meant to refer to 'eco'nomic as well as 'eco'logical priorities. Ecodyfi began their programme of integrated eco-development in the region through a public consultation process. This consultation process focused on what local people, politicians and businesses felt were the key sustainable development issues facing the area. Active consultation of this kind was seen as being vital in engaging the regional community so that they felt a sense of ownership over and responsibility towards the project. On the basis of this consultation, Ecodyfi established seven key regional development propositions (see Table 6.1). At the

Table 6.1 Propositions for regional development in the Dyfi Valley region (source: Ecodyfi 2002: 8)

1	Strengthening the local economy in order to meet social needs.
2	Prioritizing investment in tourism and agriculture as key local industries.
3	Diversification of the local economy – particularly into renewable energy services.
4	Promotion of sustainable methods in both tourism and agricultural industries.
5	Generation of distinctive local image through production of local foodstuffs and promotion of socio-cultural events.
6	Promotion of local products – including food, holidays and renewable energy innovations – as part of green image of the whole valley.
7	Promote a system of relocalization, whereby local companies and communities attempt to buy and use goods produced locally.

heart of this vision was a desire to look at ways in which economic development could be used to enhance social and environmental well-being in the area. In this context, Ecodyfi is now prioritizing economic activities related to the tourist industry and green industrial development. In relation to the development of tourism in the area, Ecodyfi is hoping to use the unique riparian and mountain environments associated with the valley as a way of attracting tourist investment and skilled service sector jobs. It is also intended that investment gained from tourism can be used to support investment in key public services, like public transport, and continued environmental protection. Ecodyfi's desire to encourage green business initiatives in the region is intended to bring skilled technical jobs to the area (particularly within renewable energy research), while consolidating the green reputation of the valley.

A whole series of programmes are now running in the Dyfi Valley region which are trying to promote Ecodyfi's development visions. These schemes include tourist provider's networks, renewable energy business promotion schemes, waste reuse projects, Welsh language programmes, local food chains and the promotion of an integrated and sustainable transport network (see Ecodyfi 2002). The exciting and varied activities currently going on in the Dyfi Valley reflect one example of how local people can construct their own regional spaces and associated modes of sustainable development. What is important, geographically speaking, about what is going on in the valley though, are the links which are being made between the spatial form of the region and its potential to be sustainable. In this context, Ecodyfi appear to be arguing that the key to creating sustainable spaces, like regions, is that they should make sense to the people living and engaging with the environment there. As we will see in the following section, however, not all visions of sustainable regional development are necessarily geared towards local community priorities.

Sustainability and the regional planning movement: the case of the West Midlands region (UK)

A sustainable region for the West Midlands

The second example of sustainable regional development I want to focus on involves a very different type of region to that we encountered in the Dyfi Valley. The region in question is actually my home region of the West Midlands (in the UK) (Whitehead 2003c). This is a very different form of region to the Dyfi Valley region for two main reasons: first because it is a much larger region; and second because it is a region which has been constructed around the large conurbation of Birmingham and the Black Country (Figure 6.6). I want to begin discussions of sustainable regional planning in the West Midlands in 2000, the year in which the first sustainable development plan for the region was devised (Government Office for the West Midlands 2000). The production of the sustainable regional development strategy in the West Midlands at this time actually reflects the growing prioritization of the region as a focus for sustainable development planning by the UK state and other Western European governments (see

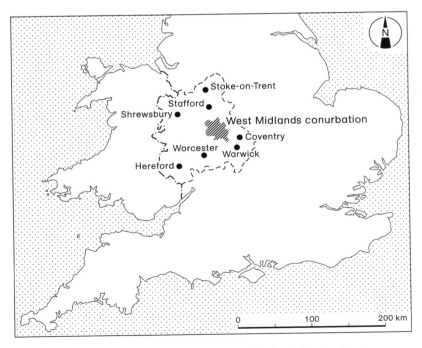

Figure 6.6 The West Midlands region and conurbation (after the West Midlands Group 1948, by Ian Gulley, from Whitehead 2003c)

Gibbs 2000; Hardy and Lloyd 1994; Roberts 1994) (for a more detailed discussion of sustainable development and the regional scale in the West Midlands see Whitehead 2003c).

I remember obtaining a copy of the West Midlands regional sustainable development strategy back in 2000 and being immediately struck by three things. First I was struck by the shear scope of the sustainable development strategy – covering as it did issues as wide ranging as nature conservation, health policies, planning, economic development and regional transport. Second, I noticed that despite its extensive coverage of regional issues, the document was relatively short, and thus represented more of a vision statement for the region than a planning strategy. Third, and finally, I was struck by the spatial area/scale which the strategy took to be the West Midlands region. While the choice of the standard government region of the West Midlands was perhaps no surprise, I was very much aware of how in the past very different West Midlands regions had been constructed to achieve very different planning objectives (see following section). I was consequently intrigued by what made this plan's designation of the West Midlands region particularly suitable for sustainable development there. But what does the West Midlands region's sustainable development strategy tell us about sustainable regional development?

From the very beginning of the document there is a strong emphasis on the need to balance regional economic, social and environmental demands (Government Office for the West Midlands 2000: 2; Whitehead 2003c: 238). This plan is radically different from previous plans for the region, which focused on housing development, economic planning and nature conservation, in the sense that it seeks to understand how all of these issues interrelate and need to be planned together. While the emphasis which is placed in the strategy on balancing social, economic and environmental priorities is not new to our discussions of sustainability, what the ideal of a sustainable region does (as was the case in the Dyfi Valley), is to change the spatial scope within which such factors are calculated and conceived. While smaller in spatial scope to a national community, the West Midlands region incorporates a much broader set of environments, economic relations and people than those associated with local or even urban communities. In this context it is clear to see that the regional community envisaged in the West Midlands sustainable development strategy is a mix of urban and rural spaces, professional elites and community groups, industrialists and agriculturalists. In spatial terms the sustainable region described within the strategy is a large-scale community incorporating a range of urban industrial districts (including the West Midlands conurbation and North Staffordshire) and rural counties (including Shropshire, Herefordshire, Warwickshire and Worcestershire), and cutting across a range of local cultural communities and economic districts (see Figure 6.6). In this context, we see a very different form of sustainable region being constructed in the West Midlands to the rural region mapped out in the Dyfi Valley. The urban–rural region being constructed in the West Midlands is much more typical of regional planning philosophies than bioregionalism. The incorporation of urban and rural districts into regional space is one of the key reasons why regions are often seen as useful planning devices when trying to balance economic, social and environmental considerations.

Beyond its scale, what is also striking about the plan for sustainable regional development in the West Midlands is the way in which it situates development within a broad range of ethical considerations (see Whitehead 2003c: 238–239). These ethical considerations are outlined in the sustainable regional development strategy in an intriguing series of questions. These questions are targeted at business people, politicians and residents of the region and essentially encourage them to think about their actions in a broader regional context (ibid.). In relation to regional businesses, it asks 'does your business location make it easy for staff to walk or cycle to work?', while enquiring of local planners whether their 'planning policy makes sure that new wildlife habitats are created and existing ones enhanced?' (Government Office for the West Midlands 2000: 8–9). Both of these questions seek to protect the regional environment through more careful planning and decision making. Other questions, however, not only seek to place local decision making within a broader regional context, but also seek to position regional decision making within a more global context. The regional plan thus asks regional citizens 'if their choice of transport reduces pollution in the atmosphere?' (ibid.). In the context of these questions then, it becomes possible to identify two key characteristics of the sustainable region. First, sustainable regional development attempts to instigate a more environmentally, socially and economically efficient use of space through the better integration of regional activities (which for example could involve the reuse of one regional industry's waste products in another regional industry's production process; reducing the need for private transport; and protecting the regional environmental systems). Significantly, these are the types of activities which it is very difficult to integrate at a local or even urban level, because of the more limited number of businesses and planning authorities which exist at this scale. Second, the sustainable region is not just an inward-looking region, but also requires regional business leaders, decision makers and citizens to understand the impacts of their region on the broader global community (see Whitehead 2003c: 239). Crucially, it is argued that both of these goals – goals of internal and external regional efficiency and justice – can be better managed, measured and understood through regional planning.

The sustainable region and the legacies of the regional planning movement in the West Midlands

While the idea of creating a more sustainable form of development in the West Midlands through regional planning may seem novel, closer inspection actually reveals a much longer history of regional planning in the area. Indeed a closer look at regional planning in the West Midlands reveals how the contemporary notion of a sustainable region has been influenced by the planning legacies and ideals of the regional planning movement (see Whitehead 2003c; Whitehead *et al.* 2006). In this context, it is important to position contemporary sustainable regional development plans for the West Midlands region in the broader historical context of the activities of the West Midlands Group on Post-War Reconstruction and Planning. The West Midlands Group (as it is normally referred to) was a collection of key regional individuals (drawn from the political, economic and civic society in the region) who lived in, or had an active concern for, the West Midlands (West Midlands Group 1948). In particular, they were concerned with the problems which were being generated by the rapid and unplanned expansion of the Birmingham–Black Country conurbation. The Birmingham–Black Country conurbation, which lies at the centre of the contemporary West Midlands region (see Figure 6.6), was one of the largest industrial conurbations in Western Europe and was one of the key geographical centres of Britain's industrial revolution (Whitehead *et al.* 2006). According to the West Midlands Group, unregulated development in the conurbation had created a series of social, economic and environmental problems. Socially, rapid industrial expansion had created a heavily polluted and unhealthy conurbation, where homes were often located near to noxious industrial plants. Economically, the unco-ordinated development of the region had generated a congested metropolis, which made the movement of industrial goods difficult and also left little room for the expansion of business premises (ibid.). Finally, at an environmental level, the lack of available space in the conurbation had resulted in a situation whereby the conurbation was continually expanding and consuming valuable rural and agricultural land. This expansion process not only saw the loss of productive agricultural land, but also, because of the lack of provision of urban green space in the conurbation itself, made it more and more difficult for urban citizens to escape the polluted city and enjoy clean air and open space (see West Midlands Group 1948; Whitehead 2003c).

The unplanned sprawl of industrial cities of course not only afflicted Britain in the twentieth century, but also affected a range of other areas throughout the world. It was in this context that a group of urban theorists, sociologists and planners developed what is now customarily referred to as the *regional planning movement* (see Box 6.3) (for more on the regional planning movement see Luccarelli 1995). This group of planners included in their number Patrick Geddes (of the Edinburgh Outlook Tower) and Lewis Mumford (with whom Geddes was in regular correspondence). What the regional planning movement argued was that the problems of the industrial city could not be solved within the city itself, but required a regional planning apparatus and mode of operation. The region was seen to be important in this context for two main reasons: first because it would enable

---

Box 6.3 Lewis Mumford and the regional planning movement

The regional planning movement was a loose association of planners which developed during the 1920s and 1930s in Britain and North America. The movement was inspired by key regional theorists and thinkers like Patrick Geddes, Lewis Mumford and Benton MacKaye who shared ideas and corresponded regularly during this era. The regional planning movement was essentially devoted to developing a new spatial framework within which the unfolding patterns of industrial modernity could be shaped and controlled. In particular, these planners were concerned with the social and environmental effects of rapid and unregulated urban development. In this context, the regional planning movement argued that regions could be used to reshape social existence under modern capitalism and provide a way of reconnecting urban residents with the natures upon which they depended. The region was important in this mission because as a scale it provided planners with a way of addressing the social and natural world (or urban and rural spaces) within one integrated plan. The vision of the regional planning movement was of ecological regions within which urban residents could constantly explore and immerse themselves in the natural world and within which a strong sense of regional social and ecological consciousness could be developed.

In order to carry forward the ideas of the regional planning movement, Lewis Mumford founded the Regional Planning Association of America, while Patrick Geddes went on to inspire a generation of British planners including the influential state planner Patrick Abercrombie. The influence of the regional planning movement can still be discerned in many towns, cities and regions today.

Key reading: Luccarelli, M. (1995) *Lewis Mumford and the Ecological Region: The Politics of Planning.*

planners to co-ordinate planning across the whole of the city or conurbation, by uniting the different urban authorities which existed in the metropolis; second, because it would facilitate planning not only across the city, but also beyond it, to incorporate key areas of the countryside upon which the city depended.

It is clear that the ideas of the regional planning movement inspired the West Midlands Group, because they strongly advocated the need for regional planning in and around the Birmingham–Black Country conurbation. Consequently, in their major publication – *Conurbation: A Planning Survey of Birmingham and the Black Country* (West Midlands Group 1948) (which actually includes a glowing foreword by Lewis Mumford) – the West Midland Group outlined a form of ecological regional planning for the West Midlands which mirrored the planning ideals of Geddes and Mumford. What the West Midlands Group suggested was the creation of a relatively small *ecological region*, incorporating the Birmingham–Black Country conurbation and its surrounding countryside (see Whitehead 2003c: 244) (this region is much smaller than that used by the contemporary regional sustainable development strategy). In the context of this region, the West Midlands Group asserted that new urban development should be contained within the existing built-up area of the conurbation (achieved largely through the more efficient use and planning of space), thus preserving the surrounding countryside as an agricultural and recreational asset. At the centre of the West Midlands Group's regional vision was the formation of what Geddes described as a *regional consciousness* among the urban citizens of the conurbation. This regional consciousness would not in this context be achieved through the use of an outlook tower, but by regular visits to a now more accessible countryside. The West Midland Group believed that the regular emersion of working-class urban populations in regional nature not only had clear health and psychological benefits, but also helped citizens recognize the balances which must be struck between a city and its surrounding environment if sustainable forms of development are to be achieved (Whitehead 2003c: 243; Whitehead *et al.* 2006). What is abundantly clear is that the types of imposed, rationally devised regional planning mechanisms developed in the 1940s in the West Midlands were a central moment in the emergence of the West Midlands as a region. It is also clear that these visions continue to influence the spatial scope and planning mentality of current sustainable regional development plans in the region.

Reflections on planning for a sustainable region

It is important to place contemporary attempts to develop a sustainable region in the West Midlands within a broader historical context of regional planning for three reasons. First, because it emphasizes how and why notions of sustainability have become inextricably tied to regionalism. While it is clear that the post-war plan for regional development in the West Midlands did not have the contemporary ideals of sustainable development in mind, it too was trying to make the region sustainable (if only in terms of a functioning social and economic space). But beyond this, in their various attempts to create a more sustainable region, both regional strategies use the region as a context for achieving greater political and economic co-ordination, whether it be through the location of housing and industry, preserving the countryside from urban expansion, or ensuring urban residents had the opportunity to encounter the regional countryside. As a planning device, then, the region provides a way of understanding important connections between social, economic and environmental systems in new ways and of potentially ensuring the sustainability of these connections through new modes of political and economic co-ordination.

Second, by looking at two examples of regional planning in the West Midlands, the constructed nature of regional space, which Allen *et al.* (1998) emphasize, becomes more apparent. In essence we see there is no pre-existing space called the West Midlands region. This region had to be named, drawn on a map and justified, and throughout time it is clear that different visions of this region have been developed. What then becomes apparent is that regions are constructed in order to achieve certain goals – like preventing urban expansion (in the case of the West Midlands Group). This issue leads us to the third and final point which is raised by the two regional plans we have discussed – that is, is the current designation of the region actually the most appropriate to achieve sustainable development? The concern I have here is that while the contemporary sustainable regional development strategy for the West Midlands is much more in keeping with the principles of sustainability (i.e. it encourages personal choice and empowerment; it seeks to position the region within a much broader set of global social, economic and environmental relations) than its historical predecessor, its choice of region makes far less sense. While the West Midlands Group constructed a specific West Midlands region in order to achieve particular goals, there is no evidence of this happening within the contemporary strategy. Indeed it is clear that the region now used to instigate sustainable regional planning in the West Midlands is an inherited, standardized regional space, which has been designated in order to assist with state administration and resource allocation. In this context, the contemporary West Midlands region does not follow particular environmental boundaries (as a bioregion would), or even recognize key cultural spaces – it simply represents a space of political administration. This could in part explain why research on sustainable development programmes operating at a regional level in the West Midlands reveals that regional partnerships are failing to engage effectively with key environmental interest groups (see Whitehead 2000). Many environmental action groups in the West Midlands see the contemporary regional strategies for sustainable development as only serving established political and economic needs (ibid.). The disengagement of environmental groups in the formation of such regional strategies only serves to strengthen the types of political and economic biases which are being engrained within the regional policy-making process for sustainable development (the separation of environmental and economic visions of sustainability at a regional level in England has recently been consolidated by the increasing desire of the British government to use regions as engines for improving England's global economic competitiveness). This is actually the same situation that is facing similar sustainable regional development planning initiatives throughout the UK. The key point here is that if regions are constructed spaces, should we not make sure that the regions we construct to deliver sustainable development are actually specifically designed to assist in this complex socio-economic and ecological process. At least in England, it appears that the regions that are currently being constructed mirror established political and economic power bases and consequently tend to fragment rather than consolidate the different interests which need to be brought together in order to develop a more sustainable society.

Summary

In this chapter we have seen how the region is being used as an important spatial framework within which to develop new sustainable geographies and lifestyles. As our two case studies show, however, the formation and implementation of sustainable regions is not as simple as we might expect. First, we have seen that the implementation of sustainable regions is not only about implementing sustainable policies within pre-existing, natural regions. Sustainable regional development often involves the contested construction of regional space, with different versions of a region (bioregion, ecological region, standard region) being preferred to achieve different goals. Second, our two case studies have illustrated the fact that sustainable regions can take one of at least two likely forms: regions which have either been constructed by grassroots community action (as in the case of the Dyfi Valley); or been imposed on to space by planners and politicians (as in the case of the West Midlands region).

Despite the contested nature of sustainable regions, one thing which has clearly emerged from this chapter is a realization of why regions have become so popular within contemporary discussions of sustainable development. As flexible, *meso* (or middle) scales (see Soja 2004), constituted somewhere between the local community and the nation state, regions provide useful spatial frameworks within which to address key sustainable development objectives. These objectives can range from balancing the needs of towns and their surrounding countryside, developing integrated transport and goods supply networks, to planning environmental systems like a river valley. It is for this reason that regions – and by definition the geographers who study and conceive of them – will continue to play a prominent role in the contested construction of a more sustainable society.

Suggested reading

For a wonderfully lucid and informative account of the regional planning movement and the broad philosophical principles behind regional planning see Luccarelli, M. (1995) *Lewis Mumford and the Ecological Region: The Politics of Planning*. For some fascinating insights into the mind of one of the key writers on bioregionalism see Sale, K. (1980) *Human Scale*. For a good summary of the links between green thinking and regional decentralization see Dobson, A. (ed.) (1991) *The Green Reader*: 73–83. For a review of contemporary policy debates surrounding the sustainable region see Hardy, S. and Lloyd, G. (1994) 'An impossible dream? Sustainable regional economic and environmental development', *Regional Studies*, 28: 773–780.

Suggested websites

To find out more about the Ecodyfi initiative go to: http://www.ecodyfi.org.uk/ For more examples of sustainable regional policies, and examples of sustainable regions in operation, go to the home page of the Sustainable European Region Network at: http://www.sustainable-euroregions.net/

7 SUSTAINABLE CITIES

Introduction

In a chapter provocatively entitled 'the urban challenge', the Brundtland Report observed that 'the future will be predominantly urban, and the most immediate environmental concerns of most people will be urban ones' (WCED 1987: 255). This observation reflects a growing realization that cities are rapidly becoming the dominant context for living throughout the world. Recent figures produced by the United Nations show that nearly half the world's population now live in urban areas and that within a generation that number will have increased by two and a half billion (UNCHS/UNEP 2001: 6). The estimated figure of two and a half billion people is significant because it is the same number the United Nations currently estimates are already living in cities (Whitehead 2003a: 1183). To put it another way, by the time that many of you who are reading this book as a student are claiming your well-earned pensions, the number of people living in cities throughout the world will have doubled from its current level! Recognizing humanity's increasing dependence on urban living is important for our consideration of sustainability for two main reasons. First, as the arena within which an increasing proportion of humanity will live out their lives, in the future cities will go a long way to determining how socially sustainable life will be on this planet. Whether it be in terms of employment, housing, health care provision, sanitation, water provision or food supply, it will be in cities that the pressures to meet escalating human needs will be experienced most acutely. Second, the growth of cities is a central issue within discussions of sustainability because, at a planetary level, it is likely that cities will increasingly become the main consumers of energy and resources and the main sources of pollution and waste in the twenty-first and twenty-second centuries.

In this chapter I argue that the sustainability of cities can be thought of in relation to both the *internal* workings and *external* dynamics of the city. When I refer to the internal workings of cities I am speaking of the social, economic and environmental conditions that are found within cities themselves. The external dynamics of cities refers to the impacts which cities have on their surrounding environments. At a local level, the external impact of cities can be observed in relation to their expansion into surrounding countryside areas, and their dependence on their local productive hinterlands for resources and food supplies. Beyond their local dependencies, the external dynamics of cities can be measured in relationship to their contributions to trans-boundary pollution and global forms of environmental change. To understand the sustainability of cities, it is consequently important to recognize the contribution which cities can make to sustainability at both a local and a more global level. In relation to the internal and external sustainability of cities, in this chapter you are encouraged to understand cities both as things and as collections of processes (see Harvey 1989). As things, cities can be investigated as specific places (like Birmingham, Jakarta, Rotterdam and Lima) which embody particular collections of roads, houses, places of work and political institutions. To understand cities as collections of processes (like economic transactions, social migrations, trading relations and housing markets), however, we have to take account of the concept of *urbanization* (see Harvey 1989). Put simply, urbanization refers to the collective processes which quite literally make different cities. The difference between analysing urbanization as opposed to cities, is that urbanization forces us to look beyond the particular qualities of individual cities (that London has a population of approximately seven million people for example), to consider the ways in which cities cast long social, economic and ecological shadows over communities and ecological systems which are located far beyond the political boundaries which are routinely given to cities. The ecological shadows of cities are often referred to as *ecological footprints*. While London's contemporary ecological footprint currently extends all around the world, the city (which has a surface area of approximately 627,000 hectares) requires a productive land area of approximately eighty million hectares to sustain its resource needs (this is close to the total *productive territory* of the whole of England) (see Girard *et al.* 2003: 15 and 45). To discuss the sustainability of cities consequently requires an awareness of both the role of cities as places where individual people live, and the broader processes of urbanization through which cities impact on more distant regions.

This chapter begins by exploring different ways in which geographers have interpreted the historical emergence of cities. In the context of the apparent unsustainability of industrial forms of urbanization throughout the world, the following section considers the emergence of the sustainable city as an international goal of urban planning. The final two sections then consider sustainable urban development through the use of two case studies. The first of these case studies focuses upon urban health reform in the British city of Stoke-on-Trent, the second considers the pressures of urban planning in one of the largest metropolitan areas in the world – Mexico City.

Understanding cities

For a short time the famous travel writer and author Jonathan Raban lived in my current hometown of Aberystwyth. In his classic text on urban living, *Soft City*, Raban (1974) explains why Aberystwyth is not a city. According to Raban, if you stand on top of Constitution Hill (a relatively small hill at the end of Aberystwyth's promenade), you can see the whole of Aberystwyth, its houses and shops, its roads and schools, where it begins and ends (incidentally 'urban' expansion means that this is no longer the case). What differentiates Aberystwyth from a city then, at least in Raban's mind, is that there is no equivalent vantage point from which you can observe the entirety of London or Mexico City: they are simply too large and complex for the eye and the mind to grasp from one viewing point. I begin with this rather peculiar way of defining a city because despite their long historical concern with urban forms, geographers persistently struggle to define precisely what cities are. Some writers have defined cities in terms of their associated ways of life and the metropolitan cultures they support, others in relation to their political and economic functions. However, I like Raban's definition of a city, not because it is necessarily the most effective, or indeed the most accurate, way of identifying cities, but because it reveals many of the vital characteristics which have historically become synonymous with metropolitan centres.

The first thing to notice about Raban's way of understanding cities is the issue of scale. Modern industrial cities are cities precisely because they have expanded to cover hundreds of thousands of hectares (see Figures 7.1 and 7.2). Cities are consequently not merely the commercial and administrative centres, or downtowns, which we often associate with them. They are vast, sprawling metropolitan regions, incorporating large suburban populations and other smaller urban districts. Second, the scale of modern industrial cities is indicative of the rapid and often unregulated expansion of cities caused by more and more people moving to urban areas to take advantage of the unique economic and cultural opportunities they offer. The third thing to notice about Raban's intuitive understanding of cities is that he emphasizes that cities are very difficult to visualize and consequently to understand as a whole. As complex collections of people, resources and infrastructures, stretching over thousands of square miles, cities have become increasingly difficult things to plan and to regulate (see

Boyer 1997: chapters 2–3; Pinder 2005). While the expansion of all large industrial cities has presented serious problems for planners, during the second half of the twentieth century a range of new megacities have emerged which have presented new challenges for social, economic and environmental policy making (see Hardoy and Satterthwaite 1992). These

Figure 7.1 Never-ending cities – New York

Figure 7.2 Never-ending cities – Paris

megacities represent urban expansion on a scale which has never been seen before in human history. The term megacities is customarily used to refer to cities with populations in excess of five million people. In 1999, the United Nations calculated that 41 megacities existed, with many of these cities containing well in excess of five million people (see Table 7.1). The exponential growth of megacities, and their rapid spatial spread, can make effective urban planning and service provision almost impossible.

If we adopt the relatively loose definition of cities as simply large-scale human communities, which involve the integrated use of socio-economic space, we are still left facing a difficult question. Why are cities becoming so important within the contemporary world? The key to answering this question is to first consider when cities initially started to emerge as the large-scale socio-economic communities we see all around us today. Because of their contemporary ubiquity, it may come as a surprise to realize that cities (at least as we understand them today) are a relatively recent phenomenon. It is estimated that in 1800 only 3 per cent of the world's population lived in urban districts (Berry 1990; Harvey 1996). Even with the dawn of the twentieth century, only sixteen cities had populations in excess of one million people, with most of these being in MEDCs (Harvey 1996: 403). By the year 2000, however, there were over 400 cities with populations in excess of one million (ibid.).

In recognizing that modern urbanization is predominantly a nineteenth- and twentieth-century phenomenon, Harvey argues that it is crucial to understand urban development in the broader context of the emergence

Table 7.1 The world's largest megacities (adapted from the United Nations 1999)

City	Population size (millions of people)
Tokyo (Japan)	26.4
Mexico City (Mexico)[1]	18.4
Mumbai (India)	18.0
São Paulo (Brazil)	17.8
New York (USA)	16.6
Lagos (Nigeria)	13.4
Los Angeles (USA)	13.1
Calcutta (India)	12.9
Shanghai (China)	12.9
Buenos Aires (Argentina)	12.6

Note

[1] Some of you may be surprised to see Mexico City only placed second in this table. In many other tables of the world's largest cities it is placed top. The reason for this variation is that the United Nations has a strict definition of where Mexico City's sprawling metropolis begins and ends. Often with megacities it is not clear where their outer limits lie.

of the military–industrial complex we associate with capitalism (ibid.: chapter 14). According to Harvey, then, the spatial form of cities supports the expansionist economic logics of modern capitalist society. In his collected writings on cities, Harvey reveals how, when viewed as a whole, cities (or more specifically city-regions) provide a *spatial fix* (or supportive context) for capitalism (Harvey 1985a, b, 1989). Cities according to Harvey provide a context within which the various outlets associated with economic production (factories, foundries, warehouses, etc.) could be located in close proximity to the vast armies of labour (people) who are constantly needed to work in different industrial sectors. In the contemporary, global and high-tech era, others have recognized the relative monopoly of cities in providing key technological infrastructures and services (see Graham and Marvin 2003: chapter 1). The concentration of economic infrastructure and people within cities also makes the provision of basic services like water supply, housing, sanitation and health care more easy to achieve (in other words cities are ideal contexts for what Castells (1977) refers to as *collective forms of consumption*). In addition to their economic functions, however, many argue that the defining characteristics of cities relates to the opportunities which they offer for social, political and environmental interaction (see Amin and Thrift 2002; Whitehead 2005). As intense meeting points for both social and environmental processes, many geographers consequently claim that cities provide unique contexts within which to think through and test the principles of sustainability.

What the work of Harvey (and other geographers working on urban theory, see Soja 1989, 1996; Smith 1991, for example) has shown, is that cities haven't accidentally replaced rural communities, villages and towns as the dominant modes of social habitation in the world today, but that there is a direct link between urbanization and the spatial logic of capitalism. To put it another way, when we study cities, we are not simply studying things (with particular population sizes and spatial areas), but how and why modern society organizes its space in certain ways. The uneven concentration of people and resources within industrial cities is consequently indicative of contemporary economic, political, cultural and social values. This is precisely why a concern with cities is so important in any geographical analysis of the sustainable society. If we are going to assess the geographical dynamics of sustainable development, we must surely ask whether the ways in which we organize the spaces in which we live, work and play are the most sustainable patterns of spatial organization we can adopt. It is in this context that the notion of a sustainable city has offered a challenge to the current spatial logic of the city. As opposed to offering a *spatial* fix for the needs of capital, the principles of sustainable urban development suggest the need to restructure cities so that they meet the needs of a sustainable society and environment as well as economy. As we will see, however, altering the spatial logic of urban development is a challenge fraught with difficulties.

Emergence of the sustainable city

Understanding the link between capitalism and the rise of the modern industrial city is a crucial step in beginning to analyse contemporary discussions of the sustainable city. Many writers, planners and social reformers for example have argued that the capitalist city represents the birth of the unsustainable urban forms to which the ideal of the sustainable city is a contemporary response. As vehicles for capitalist development, modern industrial cities reflect many of the problems associated with capitalist economic systems. The largely unregulated, market-led pursuit of economic profit within industrial cities seriously eroded the social and environmental sustainability of cities. Concerns over the social sustainability of industrial cities first started to emerge in large cities such as London and Paris during the nineteenth century, when social reformers raised concerns over the squalied, disease-ridden and overcrowded conditions which were being experienced by working-class communities there at the time (see Hall 1996: chapter 2; Harvey 2003b). The rapid, unplanned expansion of industrial cities has also generated a series of environmental problems. At one level these environmental problems were created by the overspilling of cities into countryside areas and the associated erosion of rural landscapes which such urban growth caused (see Luccarelli 1995). At another level, the unplanned nature of urban growth often meant that people had to travel long distances from their homes (increasingly in suburban areas) to their places of work and recreation. During the twentieth century this need to travel (or commute) was fulfilled by the motor car, which resulted in cities becoming loci for atmospheric pollution, noise and congestion.

Understanding how and why cities became unsustainable is vitally important if we are going to understand the principles which lie behind the notion of a sustainable city. The crucial thing to recognize here is that if the capitalist industrial city was a city devoted to being an *engine for economic growth*, the sustainable city is meant to be a city dedicated to being *an engine for socio-ecological change* (see here Girard *et al.* 2003: 11–12). While there have been many attempts to create more sustainable cities, stretching back as far as the nineteenth century, the first attempts to actually utilise the principles of sustainability within urban development and planning began in the 1970s. It was in fact recommendation I of the United Nations Conference on the Human Environment which first emphasized the importance of developing better systems of social and environmental management for human settlements like cities (Whitehead 2003a:

1185). In order to deliver this vision of a more sustainable pattern of urban development, in 1976 the United Nations launched its Habitat agenda to address urban social, economic and environmental problems. This initiative was bolstered in 1978 by the formation of the United Nations Centre for Human Settlements (UNCHS) (now known as UN-Habitat – United Nations Human Settlements Programme) as an institutional context to support sustainable urban development policies (see Whitehead 2003a: 1185). The first international programme to officially use the term 'sustainable city' was the United Nation's Sustainable Cities Programme operating under the aegis of the UNCHS. The Sustainable Cities Programme offers advice and funding for partnership cities throughout the world in their endeavour to deliver sustainable forms of urban development. The United Nations sustainable cities agenda really reached its climax in 1996 at the second Habitat conference which was held in Istanbul. The *Istanbul Declaration*, which was made following the second Habitat conference, established what is now commonly referred to as the *Habitat Agenda* for sustainable urban development (see Girard *et al.* 2003: introduction). The heart of this Habitat Agenda is a desire to transform cities from places used for the creation of economic profit to places used to promote sustainable social and economic development.

I want to conclude this section by looking in more detail at what sustainable forms of urban development actually entail. If we begin our exploration by looking at the United Nations Sustainable Cities Programme, the key principles which lie behind sustainable urbanization become apparent. The Sustainable Cities Programme states that a sustainable city is:

> [. . .] a city where achievements in social, economic, and physical development are made to last. A Sustainable City has a lasting supply of the natural resources on which its development depends (using them only at a level of sustainable yield). A Sustainable City maintains a lasting security from environmental hazards which may threaten development achievements (allowing only for acceptable risk).
>
> (UNCHS/UNEP 2001: 1)

Put simply, this quote reveals two important things about the vision of a sustainable city. First, it stresses that sustainable urban development, as with the principles of sustainable development more generally, requires us to think about cities not just as economic entities, but as integrated spaces of social life, economic growth and environmental relations (see Box 7.1). In this context, the vision of the sustainable city is not about stopping urban growth. The Brundtland Report for example recognized that cities are crucial vehicles for the production of jobs and the reduction of poverty

Box 7.1 Regeneration of Bilbao and the river Nirvion

The rise of sustainable development planning has seen a new set of relations emerging between industrial cities and the natural environment. Where once the exploitation of nature provided the fuel and resources for economic expansion, now the protection of nature is being use to re-image and re-sell cities as fashionable locations for a range of cultural, tourist and high-tech industries. Perhaps the most celebrated example of this changing relationship with nature is provided by the city of Bilbao in northern Spain. Bilbao's initial industrial development had serious consequences for the surrounding natural environment. While these consequences were expressed through air and terrestrial pollution, the most obvious casualty of industrialization in Bilbao was the river Nirvion. The river was utilized by many of the industries on its banks and rapidly became polluted and unhealthy. As part of the recent regeneration of Bilbao, a concerted effort has been made to clean the river Nirvion and to protect it from further pollution. One of the consequences of this action is that now the river Nirvion is one of Bilbao's central attractions and has seen the location of a range of new cultural institutions (like the Guggenheim Museum) on its banks (see Figure 7.3). In addition to new investors, the river Nirvion is now an important destination for tourists visiting Bilbao, and it serves to illustrate how a more sustainable set of urban relations with the natural world can have economic and social benefits.

Figure 7.3 Nature and urban regeneration in Bilbao, Spain – the Bilbao river estuary/river Nirvion

throughout the world (WCED 1987: chapter 9). However, the principles of sustainable urban development recognize that economic growth in cities can only continue in the long term if the social needs (good-quality housing, sanitation, etc.) are met, and the environmental limitations of cities are recognized. Second, and related to this point, this quote also emphasizes that the sustainable city is also seen as a city which only uses resources (including energy and materials) at levels at which the environment which provides these resources (the ecological footprint of the city) can recover.

In the context of such principles, sustainable urban development policies have taken a variety of different forms. Some schemes have for example focused upon the physical reorganization of urban space, so as to ensure that people can live close to their places of work and domestic services, thus reducing the intensity of their everyday energy requirements (see Figure 7.4). Other projects have encouraged the redesign of urban architecture to ensure that it is environmentally friendly and energy efficient (see Figure 7.5). Sustainable urban development initiatives in LEDCs, however, often focus on issues of urban slum restoration and the creation of a more socially sustainable city. The sustainable city is consequently now synonymous with a bewildering array of urban projects and initiatives which make simple definitions of what the sustainable city is difficult to form.

Figure 7.4 An advert to promote new developments on the Greenwich Peninsula (London), an integrated sustainable urban community developed by the New Labour government

Figure 7.5 Urban environmentally friendly architecture: the Greenhouse, Dublin

Beyond such general definitions of principles and practices, precisely what a sustainable city is and indeed whether cities can be sustainable at all remains a point of political and academic conjecture (see Girard *et al.* 2003 for a range of different opinions on sustainable urban development; see also Blowers and Pain 1999; Hall 1999; Haughton and Hunter 1994; Satterthwaite 1997, 1999). Some writers argue that cities provide the only sustainable spatial modes of organization for future generations (see Hall 2003), while others argue that the spatial logics of cities are inherently unsustainable (see Martinez-Alier 2003). Such debate means that the precise role of cities within the sustainable society remains uncertain (see Box 7.2). The remainder of this chapter considers the challenges facing the

Box 7.2 The Hall and Martinez-Alier debate – or can cities ever be sustainable?

In their recent contributions to an edited book on the United Nations Habitat Agenda *The Human Sustainable City* (Girard *et al.* 2003) Peter Hall (2003) and Joan Martinez-Alier (2003) captured the contemporary debates which surround the idea of the sustainable city. Building on his earlier work (Hall 1999), Peter Hall argues that cities offer the best opportunity for sustainable spatial development. Hall claims that the spatial cramming of people, resources and facilities typical of cities need not be interpreted as a cause of wasteful congestion and unhealthy living. Instead Hall argues that properly planned cities provide a context within which expanding populations can collectively receive the services and utilities which they need at least cost to suppliers, while also being able to closely integrate home, work and recreational life. According to Hall, the expansion of cities (or suburbanization) provides the opportunity to develop more sustainable, medium-density urban districts. In this context, Hall argues that properly planned cities can reduce the amount of energy needed to move people between home and work, while enabling the effective provision of often much needed socio-medical services on a large scale. Joan Martinez-Alier is highly critical of Hall's interpretation of the sustainable city. Martinez-Alier argues that the rapid rate of contemporary urban expansion means that urban sprawl continues to proceed in an unplanned and unsustainable way. According to Martinez-Alier, the increasing separation of people's homes from their places of work, as part of unplanned urban expansion, means that cities continue to witness the wasteful use of energy within transportation, while the congestion which such enforced commuting produces means that cities remain the cause, not the solution, to environmental pollution. In the context of this ongoing debate, the question remains as to whether cities can ever be made to serve the sustainable development agenda or if they will simply continue to service the expansionist needs of global capitalism.

Key reading: Hall, P. (2003) 'The sustainable city in an Age of Globalization', and Martinez-Alier, J. (2003) 'Urban sustainability and environmental conflict', in Girald *et al.* (eds) *The Human Sustainable City*.

development of more sustainable modes of urban development through two case studies of sustainable urban development in practice. These case studies illustrate the ways in which sustainable development challenges conventional understandings of cities and urban policies and the barriers which continue to confront programmes for sustainable urban development.

Sustainable city I: sick cities and sustainable health

Understanding the unhealthy city

The first story of sustainable urban development I want to consider starts in 1990 in the large urban conurbation of Stoke-on-Trent (for a more extensive account of this case study see Whitehead 2003a). Stoke-on-Trent is an English city located in the north Midlands and it is a place I am very familiar with, having carried out extensive research with local government officials, community leaders and local residents involved in sustainable urban development planning in the town (see Whitehead 2003a). Since 1990 health has dominated debates about sustainable urban development in the conurbation. The reason for this is that in 1990 a research survey of the town revealed that Stoke-on-Trent had one of the worst health records in the country, with elevated levels of a range of illnesses and diseases and a chronic problem of worker absenteeism because of ill-health (*The Sentinel* 2000a; Whitehead 2003a). This shocking insight into the health of the town was reinforced by health data produced as part of the 2001 national census which showed that Stoke-on-Trent has much higher rates of heart disease, strokes and cancer than others areas in its census region.[1]

Perhaps unsurprisingly, given its poor health statistics, Stoke-on-Trent rapidly started to be referred to as 'sick city'. While the notion that Stoke-on-Trent was an inherently sick city was obviously a concern for local government and health authorities in the area (simply because of the social suffering such poor health was creating), these health problems were gradually interpreted as part of a deeper crisis facing the town (Whitehead 2003a). Increasingly, political and economic leaders, supported by the local media, argued that Stoke-on-Trent's poor health record was fundamentally unsustainable. It was unsustainable they argued not simply because of the clear social problems it was creating, but also because of the severe economic consequences associated with Stoke-on-Trent's ill-health record. At one level the poor health record of Stoke-on-Trent was believed to be having an adverse affect on the local economy because of the high levels of worker absenteeism associated with poor health. High levels of employee absenteeism it was argued were creating an inefficient urban economy in the city and making it difficult for local industries to compete within the national and global marketplace (North Staffordshire Health Authority 1999; Whitehead 2003a). Beyond its effects on industries and businesses already located in Stoke-on-Trent, it was also claimed that the city's image, as an unhealthy place, with an unreliable workforce, was making it increasingly difficult to attract new industrial investment to the area (*The Sentinel* 2000b).

In response to the social and economic unsustainability of Stoke-on-Trent's health record, a local Health Alliance was formed in November 1993. Stoke-on-Trent's Health Alliance brought together representatives from a range of private and public bodies in the city, in an attempt to try and develop an integrated response to the health crisis (see Figure 7.6). One of the first tasks of the alliance was to investigate why Stoke-on-Trent had such a poor health record and to try and isolate the different factors which contributed to elevated sickness levels in the city. The Health Alliance quickly discovered that a model for understanding the causes of ill-health and sickness already existed. At this time, the dominant model of health being promoted by the national Conservative government was a classical neo-liberal conceptualization of sickness. On one of my many visits to Stoke-on-Trent an engaging Health Authority programme leader described this understanding of health in the following terms:

> from the late 1970s [. . .] government ideology, philosophy, thinking about health has been set by the medical agenda [. . .] if you look at right wing politics, Conservatism, it's about free

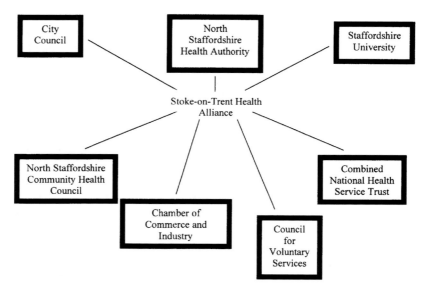

Figure 7.6 The Stoke-on-Trent Health Alliance

choice, liberty, individualism and so on [it was about] these symp-
toms not the causes [of ill health], so it was about targets set in
a very medical way, i.e. reducing coronary heart disease, reducing
cancer and so on. It was very much about lifestyle – what causes
people to die is their lifestyle, their lifestyle choices.

> (Project leader, North Staffordshire Health Authority,
> 1999, quoted in Whitehead 2003a: 1195)

The prevailing neo-liberal assumptions that ill-health was a lifestyle
choice, obviously had important implications for health policy. The Health
Alliance programme leader explained these to me:

> So basically it was seen that you have to get people to make
> healthy choices and *victim blaming* – if you choose to smoke in
> the face of all the information given about why you shouldn't, it's
> your fault. Therefore we shouldn't do anything for them, they
> have made their choice. This of course completely ignores why
> people choose to smoke in the first place. There is a social class
> gradient to smoking, as there is for every cause of death. People
> who are at the bottom of the pile die earlier, get it more often and
> smoke more than the people at the top of the pile, but we haven't
> been able to address that agenda over the last 18 to 19 years.

> (Project leader, North Staffordshire Health Authority,
> 1999, emphasis added)

Essentially the Health Alliance was frustrated with the existing models
which could be used to explain poor health in the city. According to these
models, the only reason that Stoke-on-Trent had a concentration of people
with poor health records was because it was home to a large number of
people who simply chose to live unhealthy lives. To health officials in
Stoke-on-Trent this was absurd, and they consequently explored other more
effective ways of understanding the factors which had created the city's
health problems. Eventually the Health Alliance developed what they
termed a *social model* for understanding health in the city (Whitehead
2003a). What is particularly interesting about this model in the context of
this book are its direct links with sustainable development thinking and
philosophy. The crucial thing to realize about the Health Alliance's social
model of health was that rather than conceive of health as a matter of
personal choice and lifestyle, they argued for an integrated understanding
of what causes ill-heath. The Health Alliance argued that health is a
product of a broad range of economic, social and environmental factors.
In this context, the alliance claimed that Stoke-on-Trent's poor health
record was the outcome of a large concentration of unhealthy work and

domestic environments within which many working-class people in Stoke lived out their lives. But of course poorly ventilated workplaces and badly built housing were not conditions which the low-paid residents of Stoke's industries chose to endure – they were things which their social position made unavoidable.

When you stop and consider the issues of health and sickness, it becomes clear that bodily health is one of the most basic factors which determine the sustainability of a city (see Satterthwaite 1997). Without a healthy population, it is difficult to imagine any city being socially and economically sustainable. But what makes the Stoke Health Alliance's vision of health sustainable are the ways it links the social, economic and environmental determinants of health. Referring back to the preliminary principles of sustainability that were set out in Chapter 1, it was argued that thinking sustainably involves thinking about the connections which exist between the social, economic and environmental realms of existence. In essence it was argued that things could only be sustainable if sustainability is achieved in each of these diverse realms. This relational vision of the world matches the complex model of health developed by the Health Alliance. Two consequences derive from this approach to health. First, it becomes apparent that making people healthy does not just involve supplying better health care and medication (although this is obviously vital). Such medical strategies must also be supported by economic and environmental reforms, which ensure that good health is something which is sustained well beyond an initial treatment. The second consequence of the alliance's model of social health is that health policy suddenly becomes more complex and broad ranging in its scope.

Implementing health reforms in the sick city

The radical model of health proposed by Stoke's Health Alliance initially encountered serious forms of resistance. This resistance was both political and economic in form. Initially, the Health Alliance failed to gain political support from the then Conservative government (Whitehead 2003a). The lack of central government support for the social model of health proposed in Stoke-on-Trent was partly because this model questioned the prevailing neo-liberal values of the state. It was also a product of the fact that Stoke was not eligible for regional aid monies which could have been used to finance a series of new health programmes envisaged by the Health Alliance. If the central government offered political opposition to health reform in the city, it was local employers who provided the main source of economic resistance to health reform in Stoke. The Health Alliance tried to enlist local employers into their health reform programme. Because the poor health record of Stoke was seen to be partly a product of the poor-quality working conditions found in the city, the Health Alliance felt that any programme of health improvement had to involve environmental reform within the workplace environments of the city (ibid.). The problem the Health Alliance encountered, however, was that long-term economic decline in Stoke made it difficult for employers to commit extra financial resources to health and safety initiatives in the workplace. While the ceramics and pottery industries of Stoke-on-Trent had been in long-term economic decline, the 1990s gave rise to a new wave of economic competition from corporations in south-east Asia. In the context of global competition, it became imperative for many of Stoke's leading manufacturers to keep production costs low. In other words, making commitments to invest in workplace environmental improvement was not seen as a sustainable option to many of Stoke-on-Trent's employers.

Despite the political and economic resistance experienced by the Health Alliance during the late 1990s, attitudes towards ill-health and its causes began to change. Central to this change was the election of the New Labour government in 1997. New Labour's revised brand of social democracy was immediately more sympathetic to the social model of health developed by the Health Alliance than the previous government's values had been. This sympathy was perhaps most clearly expressed in the publication of New Labour's White Paper on health – *Saving Lives: Our Healthier Nation* – which emphasized the need to tackle the broader socio-economic factors which contribute to poor health in the UK. But this backing was also a product of New Labour's desire to create more sustainable urban communities and to stem the flow of people away from inner-city areas to suburban districts (see Johnstone and Whitehead 2004: chapter 1). In this context, Stoke's Health Alliance gradually started to receive financial support for its health initiatives through the government's health and urban policy networks. Beyond the support it received from the national government, in 1998, Stoke was also designated as a member of the World Health Organization's (WHO) Healthy Cities Programme. With financial and technical support from the British government and the WHO, the Health Alliance put in place an integrated, cross-city programme of health reform called the Healthy City initiative. This initiative brings together a range of community organizations and public service providers in the city in an attempt to create more healthy living spaces. The creation of healthy living spaces is in part about trying to implement better health care and treatment through new community clinics, but it also focuses on issues of housing reform, health and safety at work, education and recreation opportunities in an attempt to develop an integrated and holistic approach to health. In this context, the attempts made to transform Stoke-on-Trent from an unsustainable, sick city to a sustainable healthy city have fundamentally been about approaching the city as a complex space of socio-economic and environmental processes, and trying to tackle poor health by addressing each of these issues together. While the impacts of the Health Alliance and related programmes are difficult to assess at this relatively early stage, figures from the latest census show some improvement in Stoke's health record. It appears that sickness rates in Stoke are moving nearer to British averages, while the illness levels in Stoke's most deprived districts are also being improved. The extent to which these changes are related directly to policies for sustainable urban health is of course open to debate.

Despite the eventual success of the Stoke-on-Trent Health Alliance in implementing a sustainable programme of urban health reform in the city, the story of Stoke provides us with a salutary lesson in the politics of sustainable urban development. Despite apparently widespread political support for the principles of sustainable development internationally, it is clear in the case of Stoke-on-Trent that the practices associated with particular brands of sustainability can encounter severe resistance when they clash with certain political and economic values. Cities are essentially spatial concentrations of social, political and economic power, and often the principles of sustainable development can be seen to challenge dominant political and economic orthodoxies in urban space. In the case of Stoke, the Health Alliance questioned how the city had been spatially constructed – from the home to the workplace, and from the town hall to the working-class housing estate. Of course, the whole purpose of this book is to emphasize the ways in which sustainable development challenges the spatial organization of the world around us. But, it is important to recognize that when it comes to cities, the creation of a more sustainable urban area cannot be simply, or unproblematically, applied to metropolitan areas. It always involves the *creative destruction* and reworking of existing spatial structures and practices – a process that will always be challenged by those who stand to lose most from such forms of change.

Sustainable city II: sustainable transportation in Mexico City

Understanding Mexico City

If you ever have the opportunity to visit Mexico City you will find yourself in a city which has fascinated and perplexed urban theorists and planners in almost equal measure. While estimates of its precise population size varies, United Nations figures put Mexico City's resident population at 18.4 million people at the turn of the last millennium (United Nations 1999). While this figure covers the metropolitan district of Mexico City, it is clear that the urban region which centres around Mexico City has a population far in excess of twenty million people. Rural depopulation in Mexico has seen Mexico City's population increase at a rapid rate, rising from 1.6 million to 13.9 million people in one thirty-year period (from 1950 to 1980) alone (United Nations, 1999). One of the main reasons for the rapid and concentrated growth of Mexico's urban population in Mexico City has been the state policies which have favoured the concentration of industry in the city over the last fifty years. Owing to its favourable location, relative to available energy sources, the Mexican state saw Mexico City as the ideal site through which it could develop a spatial economy which would transform the fortunes of the nation. Consequently, through road building and sanitation projects, and various forms of incentive, the Mexican government sought to make Mexico City the spatial hub for its new economy. The spatial concentration of people, resources, industries and infrastructure in Mexico City is really why it is so fascinating and perplexing to urban theorists.

At one level, Mexico City has been studied as an exciting and vibrant city, where different cultural values and economic opportunities meet and interact (Massey 2000). But beneath this veneer, the rapid and often unregulated post-war expansion of Mexico City has created a fundamentally unsustainable space in which to live. Mexico City is now no ordinary city, it is one of a series of urban areas which are called *megacities*. As a megacity – or city with an unusually large and/or rapidly expanding urban population – Mexico City is experiencing a series of serious social, economic and environmental problems. As the key centre of Mexico's economy, Mexico City continues to receive more migrants than it can effectively provide for. One of the key social consequences of this population explosion has been an inadequate provision of key facilities and services in Mexico City. This shortfall in service provision has perhaps been experienced most acutely in the housing sector. Increases in the metropolitan population of Mexico City mean that more than half the city's housing stock is comprised of rapidly – often illegally – constructed shanty dwellings (Pile 2000). Such settlements are often unplanned and lack proper sanitary services and transport connections. Their growth on the edges of Mexico City has seen the city expand in its spatial scope. Latest estimates claim that Mexico City constitutes a continuous built-up area of forty kilometres in diameter.[2]

The largely unregulated spatial expansion of Mexico City has combined with the rising population of the city to produce one of the city's most troubling and persistent problems – air pollution. It is currently estimated that there are 2.6 million private motorcars in the city and this figure is increasing all the time (ibid.). The expanding spatial scope of the city means that more and more people are becoming dependent upon motor cars to cross and traverse the city. The use of more cars, more of the time, is generating more and more congestion in Mexico City, which is in turn contributing significantly to the elevation of air pollution levels in the city.[3] It is now claimed that private motor cars are responsible for up to 80 per cent of Mexico's air pollution (ibid.). The remainder of this section explores how and why the spatial form of Mexico City is making it an unsustainable city, and a series of measures which are being put in place to combat the social, economic and environmental problems of the city.

Mobility and the workings of a city: the problems of the congested metropolis

As we have already discussed, one of the key problems with urban development in Mexico City is congestion. A rapid expansion in the city's population has combined with an unregulated pattern of spatial development to make it very difficult to move around the city. Issues of mobility – or how people move (or don't move) through space – are important concerns for the geographer (Cresswell 2001, 2006). When you actually stop and think about it, movement is at the centre of how our societies, economies and ecosystems operate. Whether it is the movement of workers from their homes to their jobs, raw materials from mines and excavation sites to factories and power stations, or nutrients from the soil into plants and eventually into the human digestive system, movement quite simply makes life possible and sustainable. In his excellent work on the role of mobility and movement in social, economic and cultural history, Tim Cresswell (2001, 2006) illustrates how motion is central to human existence. In a specifically urban sense, popular discourses emphasize the importance of keeping a city *on the move*, or preventing an urban centre suffering the stagnating effects of becoming *clogged up*. Mexico City's problems lie in the fact that while recent developments have necessitated the need for more personal mobility (because of urban spatial expansion), this opportunity for personal mobility has been blocked by more people needing to move (a product of an expanding urban population).

We all encounter the effects of congestion – whether it is the delayed bus, the traffic jam or the belated delivery of that crucial package we are waiting for. In this context, congestion seems like an inevitable part of modern life and as such is something which I tend to pay relatively little critical attention to. But when you actually stop and think about it, issues of congestion, movement and mobility are crucial considerations in any discussion of the geographies of the unsustainable society. In Mexico City – where admittedly it has reached unusual levels – urban congestion is not simply an inconvenience, it is a stubborn barrier to sustainable development. At a social level, it is congestion which is making it difficult to provide affordable housing areas which are close to the places where working-class communities work. Furthermore, congestion on the roads and streets of Mexico City, as we have already seen, is generating high levels of air pollution, which is in turn contributing adversely to human health in the metropolis. Air pollution has been linked to elevated levels of childhood asthma and numerous other respiratory illnesses among the adult population. In Mexico City, air pollution quite simply makes people ill – it is socially unsustainable. Beyond the social effects of environmental pollution, it is also clear that Mexico City is a major global producer of greenhouse gases. In this context, it is clear that the car-dependent culture of Mexico City has consequences for environmental sustainability well beyond its metropolitan boundaries.

In addition to the social and environmental costs of congestion in Mexico City, congestion has also created serious economic problems in the metropolis. With constant delays on Mexico City's transport network, the efficient movement of economic goods and resources becomes difficult and businesses in the city can be greatly disadvantaged. Beyond the potential impact of delays on economic profit margins, however, the image of Mexico City as a heavily congested, polluted and essentially inefficient city has also made it difficult to attract new investment and important multinational corporations to the area. Mexico City's congestion problems are undermining the economic competitiveness of the whole metropolis. When you remember the importance of Mexico City as a growth pole for national economic development in Mexico, the unsustainable nature of urban development in the city is given even greater significance.

Strategies for sustainable urban transport in Mexico City

There are obviously numerous ways in which different city authorities have sought to tackle issues of urban transport congestion. What many of these strategies have in common – whether it be road tolls, congestion charges or increased investment in public transport – is that they don't approach issues of congestion in an integrated way; that is to say that they don't consider the implications of congestion from the perspective of sustainable urban development. In Mexico City, for example, a scheme was launched in 1989 which attempted to reduce traffic congestion by restricting the days upon which private cars could be used; this process was determined and monitored on the basis of car licence plates (Eskeland 1992). Analyses of this scheme have been highly critical, claiming that the project failed to recognize the different social and economic needs of people living in the city, while not addressing the underlying causes of increased private car use and congestion (Eskeland 1992). The key to sustainable urban transportation and development, according to writers like Peter Hall (2003), is to develop effective planning systems for the whole of the metropolitan area. In other words, achieving a sustainable urban transport system is not something which can be achieved by transport departments and planners alone, but requires the transformation of the whole spatial economy of the city. Recent developments in Mexico City have seen transport policy recast in these more holistic sustainable terms.

The origins of sustainable transport planning in Mexico City began in 2002, when the ministries of Transportation and Environment for Mexico City began working with the World Resources Institute (Box 7.3) on a scheme to implement sustainable solutions to Mexico's transportation problems (World Resources Institute 2002). The result of this partnership was the creation of the Center for Sustainable Transport in Mexico City (hereafter CST). The CST is a non-governmental organization which works in partnership with Mexico City's metropolitan authorities, the World Resources Institute, the World Bank (and in particular the Bank's Global Environmental Facility) and major motor vehicle and fuel suppliers to

---

Box 7.3 World Resources Institute

The World Resources Institute was established in 1982 through the financial support of the Catherine T. MacArthur Foundation. It was created to provide policy research and evaluation for resource management and environmental policies. The founders of the World Resources Institute felt it was important that as an organization it be non-governmental and not for profit. This independent status, it was felt, would help the institute gain local support for its social and environmental programmes in a wide range of countries. Utilizing over 100 scientific, economic and public policy experts, the World Resources Institute delivers research which can be used by public and private sector partners in different areas to promote social and environmental reform. In this context, the institute does not simply offer development funding to local projects, but through its research programmes it also seeks to create strategies which other partners can become involved with and offer financial support to. In this context, it is interesting to note that the Center for Sustainable Transport in Mexico City was partly financed by the Shell Foundation. As a large oil manufacturer, many have become suspicious of Shell's involvement in the CST and the wider motives of the scheme in general.

Key reading: World Resources Institute's web page: http://about.wri.org/

tackle the problems of traffic congestion and air pollution in the city (World Resources Institute 2002). The main goals of the CST are threefold: (1) to reform public transport systems in the city which are seen as being overcrowded and unattractive to commuters; (2) to encourage fuel and efficiency modifications to many of Mexico's ageing and inefficient private cars; (3) to develop systems of land-use planning which reduce transport pressures and the need to travel (World Resources Institute 2002).

If you read the policy documentation associated with the CST it is clear that the CST is trying to understand transport as a social, economic and environmental issue. In this context, a background paper on the centre published by the World Resources Institute states:

> Business, civic, and government leaders in Mexico City realize the pressing need to improve public health, increase community prosperity, and change the image of Mexico City. They recognize that the image of widespread traffic congestion and almost constant air pollution impedes the city's ability to attract investment, tourists, and skilled workers necessary to increase the region's economic opportunities and improve the overall quality of life.
>
> (World Resources Institute 2002)

This quote reveals a growing realization in Mexico of the necessary complexities associated with transport planning and how transport issues have implications for Mexico City at a range of different scales (including local pollution and global trade). The recognition now given to the broad set of socio-environmental problems associated with Mexico City's transport system has also involved a fundamental transformation in the ways in which such problems are now being tackled. Again turning to the World Resources Institute:

> The long term goals for Mexico City's transport policy are to reduce travel times and costs, improve the economic and environmental performance of each mode of transport (including private cars), restore a balance of modes re-emphasizing metro and buses, and integrating the various systems. A mix of policies involving vehicle and fuels standards, land use controls, prioritizing bus systems, and economic stimuli is required. This will require actions at the city, regional and, in some instances, at the national levels of government.
>
> (World Resources Institute 2002)

This *mixed-approach* to transport planning is a typical feature of contemporary sustainable transport policy. Such an approach is based upon the realization that both the causes and resultant problems created by inefficient transport systems are not simply products of the transport system itself.

Although the CST is still in its infancy, it has implemented a series of sustainable transport schemes. Particular emphasis, as I have already mentioned, has been given by the CST to the reform of public transport in the city. In an attempt to develop a greater mix in the modes of transportation used in Mexico City, and in particular to promote mass-transit systems in preference to private motor cars, the CST has promoted transport reform on the city's metro and bus service. One of the major forms of public transport reform instigated by the CST has been the introduction of a bus rapid transport (BRT) system, which provides priority lanes for buses so they can avoid congestion. The CST has also been actively involved in promoting the Mexico City Diesel Retrofit Project. This project, which is supported by the US Environmental Protection Agency, uses retrofit technologies to reduce the amount of air pollution generated by Mexico City's fleet of diesel buses (Environmental Protection Agency 2004). It appears that although the CST project is having some success in the technical aspects of its remit – namely improving vehicle efficiency and developing alternative transport networks – it is still proving difficult to harmonize urban planning in Mexico City in a way that would benefit sustainable transport. As a city that is still rapidly expanding, Mexico City, it would appear, continues to meet the demands of modern capitalism first and sustainability second.

Although many are still unsure as to the effectiveness and motives behind the CST in Mexico City, it is clear that the centre is trying to think about urban mobility in sustainable terms. While it is unlikely that the types of transport policies being developed by the CST are going to address the culture of car usage which lies at the heart of Mexico City's (and other major cities') contemporary congestion problems, its policies do emphasize the need for a more holistic approach to transport policy. Furthermore, in terms of this book, the work of the CST in Mexico City also emphasizes that the geography of the sustainable society is not simply a question of static spaces and locations, but is also an issue of how social, economic and environmental things/processes move through space and between places. A geographical analysis of the sustainable society should consequently be just as much concerned with how spaces are connected as it is with what goes on within those particular spaces.

Summary

In this chapter, we have explored how and why cities are becoming such a dominant way of organizing social, economic and environmental space throughout the world. In the context of the emergence of urban living worldwide, this chapter has argued that any discussion of the geographies of the sustainable society must consider the spatiality of cities. As geographers with a concern for issues of sustainability, when we look at cities we must consider the implications of this form of spatial development for wider forms of sustainable development. As we have seen, some (for example Hall 2003) argue that the spatial form of the city, as a concentration of people and resources, is actually, when appropriately planned, a good *spatial fix* for the sustainable society. Others have argued that the congestion and unplanned spatial expansion of cities makes them a fundamentally unsustainable form of socio-economic development (for example Martinez-Alier 2003).

Beyond the precise role of cities within the sustainable society (that is as vehicles or obstacles for sustainable development), the two case studies we have considered have illustrated how notions of sustainability can change the way we understand cities and address urban problems. In the case of Stoke-on-Trent, we have seen how poor health in the city was preventing sustainable forms of development in the metropolis. Furthermore, we saw how narrowly conceived neo-liberal understandings of Stoke's poor health record were failing to address the problems of the city. By exploring the policies of Stoke's Health Alliance and the subsequent Healthy City Programme, we have seen how notions of sustainability (in this case expressed through a social model of health) force us to consider the broader social, economic and environmental determinants of ill-health. Of course as soon as urban health is seen as a product of these diverse forces, health policy becomes part of a wider reform of the economic practices and the physical fabric of the city. In the case of Mexico City, we have seen that it is not only the spatial form and fabric of the city, but also the ability of things to move around the spaces of cities, that determines their ability to be sustainable. In the case of Mexico City, urban population growth, the spatial expansion of the city and an increasing dependency on privatized motor-car transport have all contributed to stopping things moving in the city. Through our brief exploration of the policies of the Center for Sustainable Transport in Mexico City, we have seen the importance of developing holistic economic and social policies for dealing with transport in a sustainable way. In both the cases of Stoke-on-Trent and Mexico City, different as they are, we have seen that sustainable urban development is often very difficult to implement, but it is often the strategy which is most likely to address the root causes of urban problems.

Suggested reading

For a good overview of the causes of unsustainable development in cities and the different historical strategies which have been developed to make cities more sustainable see: Blowers, A. and Pain, K. (1999) 'The unsustainable city', in S. Pile, C. Brook and G. Mooney, (eds) *Unruly Cities?*: 247–298. For a critical analysis of the principles of sustainable urban development and more on Stoke-on-Trent's Healthy City Programme see: Whitehead, M. (2003a) '(Re)Analysing the sustainable city: nature, urbanization and the regulation of socio-environmental relations in the UK', *Urban Studies*, 40: 1183–1206. For an overview of sustainable urban development policies see: Haughton, G. and Hunter, C. (1994) *Sustainable Cities*.

Suggested websites

For more information on urban growth and megacities go to the Population Reference Bureau at: http://www.prb.org/Content/NavigationMenu/PRB/Educators/Human_Population/Urbanization2/Patterns_of_World_Urbanization1.htm

For more on the role and associated programmes of UN-Habitat see: http://www.unhabitat.org/

More information on Stoke-on-Trent's Healthy City Programme is provided at the scheme's home page at: http://www.healthycity.stoke.gov.uk/

To find out more about the goals and projects of the World Resources Institute see: http://about.wri.org/

8 LOCALIZING THE SUSTAINABLE SOCIETY

Between citizenship and community

Introduction

There is a well-known saying that *all politics is local politics*. This statement should not of course be interpreted as a claim for the supremacy of local over other forms of national or international government. Instead this simple observation serves to remind us that all politics – whether it be President Bush making decisions in the Oval Office of the White House, or the ways in which US foreign policy impinges upon different communities throughout the world – originates from, and ultimately impacts upon, localities. In the context of this book, I would like to adapt this well-known adage to claim that *all forms of sustainability are ultimately local sustainabilities*. To recognize the local nature of the sustainable society is important at this stage of this volume. So far in this book we have discussed sustainable development in different states like the United Kingdom or Kenya, in different regions like the West Midlands or the Dyfi Valley, in large cities like Mexico City and Stoke-on-Trent, and also as a global phenomenon stretching beyond individual national communities. In all these important discussions, however, there is a danger that we can forget that facets of the sustainable society are always already around us. Yes, even now as you sit reading this book, you are contributing to the processes which determine the relative sustainability of the world we live in. Ask yourself for example how you got to the library (if that is indeed where you are) in order to read this book – did you drive your car, cycle, catch a bus or walk? Where, and under what conditions, were the clothes you are currently wearing made and manufactured? What about the paper you are taking notes on, or even on which this book has been printed – has it been recycled? *All forms of sustainability are ultimately local sustainabilities.*

Recognizing the local nature of the sustainable society is important in the context of our discussions of the geographies of sustainability. Geographers have a long historical interest in localities – understood as those particular spaces where more general historical patterns and processes find geographical expression and form. More recently geographical discussions of *the local* have focused upon two crucial issues – the questions of community and citizenship. In many ways the notion of local space is synonymous with the idea of a *local community* who inhabit and share that space. This is the community of people we share our everyday living spaces with, whom we recognize and identify with, and on whom our socio-ecological actions often most directly impinge. Related to this vision of a local community is the concept of citizenship. Citizenship can be understood simply as a system of rights and responsibilities in and through which we organize our lives and actions (see Desforges 2004). As we will see later in this chapter, while modes of citizenship can operate at a variety of geographical levels (including the region and nation state), often it is at a local level that our rights and responsibilities as good citizens are realized and judged. This chapter explores the challenges which notions of sustainability are presenting to existing and entrenched geographical understandings of locality, community and citizenship. In this context, we will see that while the sustainable society depends upon local actions and new modes of citizenship, these cannot be confined to narrowly defined territorial notions of community. This chapter begins with a brief discussion of the relationship between geography and the study of local communities and citizenship. Analysis then moves on to consider the challenges which notions of sustainability present to traditional geographical interpretations of the spatial dimensions of local communities and citizenship. The final section then analyses the opportunities and barriers facing sustainable development policies by considering two case studies: one a Local Agenda 21 scheme in Sweden, the other an education programme for sustainable citizenship in Wales.

Local worlds and the geographies of citizenship

The local as a site for sustainable development

The concept of 'the local' has been a prominent concern among geographers over the last thirty years (see Duncan and Savage 1989; Massey 1993; May 1996; Murdoch and Marsden 1995). But during this time, what the local actually is has been heavily disputed and rigorously debated. While we all have some sense of what the term local means, when we attempt to define it more precisely difficulties inevitably develop. In the context of your own lives, for example, is your local space your home, or your neighbourhood? You may define your local area as a larger-scale community – perhaps an urban district or a rural county. The point is, while the notion of the local suggests a sense of geographical proximity, or closeness, we all define and interpret this proximity in different ways and at vastly different scales. In a widely cited book, entitled *Place and Politics,* John Agnew (1987) develops a framework within which to order different and competing geographical definitions of local space. Agnew argues that local spaces can be understood in relation to one of three inter-related concepts: as *locales* (understood conventionally as the social and physical setting within which various forms of human (and non-human) interaction occur); *localities* (that is to say precise places which can be located on a map through grid references); and *senses of place* (or the way in which humans attach social and cultural meaning to their local living spaces).[1] Within these different understandings of the local it is possible to discern two main ways in which local spaces can be understood: (1) as politically designated territories encompassing bounded places which are normally smaller than those occupied by cities or regions; (2) as socio-cultural contexts within which human beings routinely interact and in the process of this regularized interaction, give meaning to the world around them. In the context of the contemporary forces of globalization (see Chapter 5), many prominent geographers have started to question whether what makes a local space local can really be understood in terms of local interaction alone (see Massey 1994; May 1996). Writers like Massey (1994, 2004) have consequently sought to uncover how local spaces are actually constituted by complex webs of relations which transcend locali-ties. In this sense, what appears to make local spaces local is not so much local processes but how more general social, economic and cultural forces connect and coalesce within particular local areas.

In relation to these different interpretations and debates concerning the constitution of the local, it is perhaps unsurprising that the idea of local communities has also been keenly debated within geography. At one level perhaps the existence of local communities seems to be self-evident. We all probably have some sense of the types of people who belong to our own local communities and where those communities are located. But many geographers and social scientists have become increasingly suspi-cious of the idea of society being made up of a series of spatially bounded, discrete local communities. At one level this vision of local communities appears dangerous, suggesting as it does a clear sense of who belongs to a community and who is perhaps unwelcome within that community space (see Sennett 1974). Furthermore though, this static notion of fixed, clearly demarcated and tightly knit communities is somewhat anachronistic within the contemporary global era. Consequently, while the idea of safe and inte-grated community spaces may be appealing, it is, according to some, based upon nostalgia for a bygone era, rather than an accurate description of the contemporary organization of social space (Massey 2004). The key insights of contemporary geographical reflections on the notion of local communities have consequently been twofold: (1) a realization that visions of the 'local community' are often constructed along specific cultural and political lines, and that these visions selectively include and exclude different social and cultural groups – in this context while the notion of a local community can be reassuring and comforting for some, for others it can be a highly exclusionary and alienating experience (see Box 8.1); (2) that the pervasive forces of globalization make the idea of geographically bounded, local communities increasingly difficult to sustain – any given community is now clearly permeated by a range of social groups, economic

---

Box 8.1 In and out of place: Tim Cresswell on the dark side of local places

Tim Cresswell's pioneering work on place provides us with some salutary insights into the nature of place-based and localized policy making within the sustainable development debate. Cresswell's work has been devoted to excavating the inherently political nature of places. While places are often constructed as naturally occurring, safe and reassuring geographical havens, Cresswell illustrates that the construction of place is an inherently political act, which is informed by the pervasive power of numerous ideologies. Drawing on the example of different marginal groups – including tramps, peace protesters and graffiti artists, Cresswell shows how different people are constructed as 'in place' or 'out of place'. The designation of whether someone fits in to a place or not can be determined according to their appearance, gender, ethnic background or cultural identity, but this process of inclusion and exclusion is a highly significant moment within the designation of the local. Cresswell considers the different ways in which society creates its places and prevents unwanted transgressions into these arenas by so-called 'other' groups. In shattering many of our nostalgic assumptions surrounding notions of place and local areas, Cresswell revels the dangers which the uncritical valuation of place can create. In this context, when we talk about the development of sustainability in local places, it is important to consider who is deemed to be 'a local' (and by whom), and the alienating effects which the construction of bounded communities can have on those who are seen as being out of place. These are crucial questions when considering the localization of sustainability.

Key reading: Cresswell, T. (2004) *Place: A Short Introduction*; Cresswell, T. (1996) *In Place/Out of Place*.

products and cultural influences which originate well beyond the boundaries of the community (Massey 1994).

I introduce these debates on the complexities surrounding the idea of the local not to wantonly confuse the reader, but because the idea of the local has been routinely taken for granted with discussions of the sustainable society and associated sustainable development policies. A quick glance at the various ideologies and policies surrounding officially sanctioned sustainable development ideals reveals how central the idea of the local has been within discussions of the sustainable society. If we take the United Nations' *Agenda 21* document (the agenda for the twenty-first century) – a great proclamation on the principles of sustainable development – as an example, we find constant references being made to the need to move towards more indigenous-based and locally sensitive strategies when attempting to deliver more sustainable communities. The link between sustainability and localization clearly finds its origins in the green movement's broader critique of the inefficiencies and ecologically alienating effects of nation state-led environmental policy regimes. But the link between sustainable development and the local is undoubtedly expressed most directly in the establishment of the Local Agenda 21 programme. The Local Agenda 21 (hereafter LA21) programme was launched at the Rio Earth Summit in 1992 (UNCED). The primary aim of LA21 was to provide the mechanisms through which the visions of Agenda 21 could be delivered and applied within different local contexts. The Agenda 21 document states that:

Each local authority should enter into a dialogue with its citizens, local organizations, and private enterprises and adopt 'a local Agenda 21.' Through consultation and consensus-building, local authorities would learn from citizens and from local, civic, community, business and industrial organizations and acquire the information needed for formulating the best strategies.

(Agenda 21, Chapter 28, Section 1.3)[2]

This quote reveals a number of points concerning the link between the official discourses of sustainable development and the local. First, we can discern the long-established belief that through the localization of policies sustainability can become more sensitive to the 'real needs' of communities. The localization of sustainable development is thus supposed to enhance the process of public consultation and engagement – it is meant to actively involve people in the creation of a more sustainable society. Second, the sentiments of Agenda 21 clearly emphasize that local spaces provide important contexts within which agreements and consensus over the direction which sustainable development should take can be made.

Finally, the idea of Local Agenda 21 has a very clear, if narrow, under-standing of what the local actually is. Within the activities of LA21 schemes the local is axiomatically taken to be those pre-given, politically designated spaces we associate with local government.

The local geographies of sustainable citizenship

In the quote we have just reviewed from the Agenda 21 document there is a clear link being forged between the localization of Agenda 21 and the issue of citizenship. Citizenship has been an emerging area of debate within critical geography over the last fifteen years (see Brown 1997; Marston and Staeheli 1994; Painter and Philo 1995; Smith 1990).[3] Put simply, as an area of study citizenship is concerned with how the members of different communities are connected through complex webs of rights and responsibilities (Smith 2002: 83). Given its concern with political conduct and behaviour, the study of citizenship and citizens is clearly significant within any discussion of the sustainable society. As with any political com-munity, the sustainable society is dependent upon its constituent citizens for its very existence. It is citizens who will do all of the living and the breathing and the dying within any sustainable society, and it is the actions of citizens that determine just how sustainable a given society actually is. In the next section we will consider the challenges which the idea of the *sustainable citizen* presents to traditional modes of citizens. For the remainder of this section, however, I want to consider the links between citizenship and local space.

As essentially a set of relationships forged between an individual and a broader political community, citizenship has historically been interpreted in relation to one political institution in particular, that of the nation state (for more on the longer and pre-national history of citizenship see Isin 2002). As one of the dominant modes of political organization in the modern world, states have been interpreted as pivotal institutional contexts within which the rights and responsibilities of citizenship are expressed and realized (see Brown 1997: 85). If you stop and think about your own citizenly relations, you will notice that they continue to be heavily (if not exclusively) defined within the parameters of nation states. Your political rights to vote, free speech, health care, pensions and shelter, are secured – to admittedly very different levels – by the state you reside in. At the same time, many of your political obligations are also defined in relation-ship to that state – including military and community service, paying taxes, and various duties of care (or non-harm) to your fellow citizens. In this context it is perhaps convenient to think of citizenship as the socio-political and ethical relations forged between a state and its residents – or to express it in a more technical way, between a nation state and civil society. While this perspective provides us with a convenient way of understanding citizenship, the innovative work of Michael Brown (1997) illustrates that it can be misleading (see Box 8.2). First, Brown's work illustrates that the formation of modes of citizenship is not something which is exclusively

Box 8.2 'RePlacing citizenship' with Michael Brown

Michael Brown's work on the spatial dimensions of citizenship has become influential in geography in recent years. Focusing on the relationships which have been forged between the state and civil society over the provision of HIV/AIDS services in the Canadian city of Vancouver, Brown's work consistently emphasizes the spatial nature of citizenship. Michael Brown revealed the importance of local grassroots movements in dealing with the health and social difficulties which are faced by HIV/AIDS sufferers. Given the slow reaction of the state to HIV/AIDS, Brown highlights how local grassroots movements provided important sources of accessible welfare support to HIV/AIDS communities. Despite providing radical solutions to the problems associated with HIV/AIDS, Brown notices how many grassroots movements in Vancouver gradually became dependent on the state to support and fund their work. In this context, a situation emerged whereby the state was able to use locally (often voluntary) based grassroots service infrastructures as a basis for its own HIV/AIDS policies, while the operation of such grassroots facilities were greatly extended and formalized because of the state support they received. The question, which Brown asks, is what effect does the relationship between the state and various grassroots movements have on the nature of local citizenship? He argues that while state support can clearly enhance local welfare provision, it can also change the nature of such provision from being an act of voluntary (and perhaps radical) citizenship, to one of bureaucratic service delivery on behalf of the *shadow state*. Ultimately Brown's work stresses the need to recognize the spatial mixing of the state and civil society within emerging modes of citizenship and to explore this relationship within the increasingly complex local geographies of grassroots radicalism and civic voluntarism.

Key reading: Brown, M. (1997) *RePlacing Citizenship: AIDS Activism and Radical Democracy*; Brown, M. (1994) 'The work of city politics: citizenship through employment in the local response to AIDS', *Environment and Planning A*, 26: 873–894.

focused on, or geared towards, states. With the decline of welfare state provisions in many Western states over the last thirty years, there has been a proliferation of new political communities of self-help and mutual support which have been forged around issues of health care, the environment, community safety, and ethnic and gender exploitation. Brown asserts that these new political communities, with their own systems of care and responsibility, are increasingly being located at a distance from the formal bureaucracies of the state (see Brown 1997: chapter 4). According to Brown, these radical versions of citizenship have been forged in between the formal institutions of the state and civil society. Such radical versions of citizenship consequently reveal that while citizenship may involve relations with the state, these relations are not always centred upon the political community of the state. Brown's second and related observation is that abstract analyses of the relationships between the state and civil society often fail to recognize the local geographies which inform citizenship. Brown observes that '[S]tate and civil society do not exist in any pure abstract sense, but are bundles of social relations that are always located somewhere' (1997: 85). In making this statement, Brown calls upon those working on emerging modes of citizenship to recognize their constitution within a diverse range of local spaces. This *RePlacing* of citizenship serves to emphasize that, despite the persistent influence of nation states, citizenly relations are always constituted and expressed within particular local circumstances.

In order to explain Brown's interpretation of the relationship between state, space and citizenship, I want to reflect upon an event I observed just before I began writing this chapter. During a recent state holiday I was walking along the seafront in Aberystwyth when I noticed something which at first seemed rather peculiar. A tight cluster of people, tables and display stands occupied the bandstand, which is normally used for open-air music performances. A chalk noticeboard at the entrance to the bandstand read: *Talk on eco-friendly nappies (2pm); Discussions on local green activism (4pm)*, while inside I noticed young children making various constructions from a collection of recycled goods. After some provisional investigation I discovered that the bandstand was the hub of the 'Aber is Green' weekend (see Figure 8.1). This weekend had been co-ordinated by local green charities and voluntary organizations in conjunction with the local peace campaign. Inside the bandstand, stalls informed passers-by of how they could become more actively involved in local recycling schemes, green campaigns and the opportunities which existed locally for buying environmentally friendly and ethical products. I think that two points are most striking about the 'Aber is Green' campaign for our discussions of citizenship. First, it is clear that various voluntary sector organizations involved in the campaign are trying to forge a brand of local citizenship

Figure 8.1 'Aber is Green' and the rise of new spaces for sustainable citizenship

which is situated at a distance from the state. This is a campaign which is attempting to construct a new set of duties and obligations among local people. These new obligations stress that the civic community should no longer depend on local and national government to deal with environmental issues, but should instead take a more active role in environmental management itself. Despite being located at a *distance* from the state, closer inspection shows that many of the organizations involved in the 'Aber is Green' event did receive (or were seeking to receive) support from various local, regional and national government agencies. In this context, and as Brown emphasizes, even new, radical modes of citizenship continue to be forged in between informal and formal political structures, or between civil society and the state. The point is that as citizenship becomes localized, a negotiation process clearly takes place. This negotiation process involves the prioritization of local issues within emerging brands of citizenship and the continued need for state resources to deliver and promote these priorities.

The second important point to notice about this brief case study, is the geography of the 'Aber is Green' campaign. By locating their activities in the bandstand, the campaign really achieved two important goals. First, the campaign was able to reach out effectively to the wider community (particularly those passing along the seafront), and engage them in issues of ecological citizenship in ways which perhaps town council meetings or planning hearings never can. Second, by being located in the bandstand, the 'Aber is Green' campaign reflected a shift in the spaces where local and environmental citizenship takes place. This form of citizenship, it would appear, can no longer be confined to the town hall, voting booth or council meeting, but involves activities which permeate all aspects of everyday public and private life, from the office to the home, and from the supermarket to the classroom (see Brown 1997; Dobson 2000).

If, as I have suggested, we are to understand local citizenship as a set of rights and responsibilities which are negotiated between the state and civil society, it is important to recognize that different modes of citizenship exist which reflect different points in this negotiation process. At one level citizenship can take a fairly passive form, to the extent that it involves the uncontested acceptance of political responsibility on behalf of the citizen and the delivery of key services to the citizen by the state. In this context, a fairly inactive form of citizenship could involve the paying of taxes to the state by the citizen and the use of the welfare services which are funded through these taxes by the same citizen. These modes of citizenship are often referred to as *clientalist*, or where the citizen uses the services of the state like a client would draw on the services of a private company (see Brown 1997: chapter 4). In contrast to this form of passive citizenship, many modes of citizenship can take a much more active form. Active citizenship, as the name suggests, involves citizens taking an action-oriented approach to their communities and associated duties (Kearns 1992). Consequently, rather than simply doing what you have to do, active modes of citizenship involve people recognizing the different ways in which their skills and energies can be used to support and bolster their communities. Whether it be through neighbourhood gardening initiatives, or home-help services, the active citizen is a citizen who attempts to compensate for the gaps within the state's own welfare provisions by providing uncosted, voluntary support for their communities (Smith 2002: 83). Active citizenship can often be subservient to official state views of what citizens should do and what ethical codes they should follow. Various geographers and social commentators have charted an alternative, but still *active* brand of citizenship. This mode of citizenship is customarily referred to as radical citizenship. As the name suggests, the radical citizen is someone who questions the values and practices being promoted within dominant modes of state-centred citizenship, and seeks to promote alternative systems and spaces of citizenly relations. Perhaps the most obvious examples of radical modes of citizenship have emanated from within the environmental movement. Eco-activists have for example consistently questioned the consumption-centred, anthropocentric modes of citizenship promoted in Western democracies, and promoted alternative systems of environmental rights and responsibilities (see Figure 8.2). One of the most celebrated examples of radical environmental and sustainable citizenship

Figure 8.2 'The end of the road' – radical citizenship and environmental protests against the car culture

has been the *Environmental Justice Movement* in the USA. This group have revealed the impacts of race and gender on environmental rights and questioned what environmental politics should be about (see Box 8.3). As we see in the following section, the principles of radical environmental citizenship have had a significant impact upon emerging modes of sustainable citizenship.

Fracturing the local geographies of citizenship: the challenge of the sustainable citizen

The previous section outlined the key characteristics associated with the conceptual categories of locality and citizenship. We also saw how the emergence of sustainable development has connected the notions of locality and citizenship in new ways. In essence we have seen that the vision of a locally sensitive and active brand of sustainability is now intimately tied to the creation of more sustainable and active modes of citizenship. However, in this section I want to challenge the officially

Box 8.3 Environmental Justice Movement

Robert Bullard, who has written extensively on the Environmental Justice Movement (or EJM), argues that the movement finds its origins in a set of local political struggles in Warren County, North Carolina, during the late 1970s (Bullard 1990). In 1978, Warren County was chosen as a site for the disposal of 43,000 tons of soil which had been polluted with PCBs (PCBs – polychlorinated biphenyls – are dangerous toxic compounds which have been related to a range of human health problems) (Bullard 1990). The interesting thing about Warren County was its marginal status. As a county it had relatively low-income levels, a high percentage of people of colour among its population, and a high ratio of elderly and retired citizens. It was in the context of these demographic conditions that the people of Warren County started to believe that they were being targeted as a place for pollution disposal and treatment precisely because they were a relatively marginal and disempowered place. This concern led to a range of protests in Warren County during the late 1970s and early 1980s. Within these protests a political discourse started emerging which claimed that the environmental rights of marginal groups within American society were not being protected as strongly as those of more powerful, often white communities.

Following on from the celebrated events of Warren County, a whole series of protests started to emerge across the USA concerning the environmental conditions which marginal communities were facing on a daily basis. These protests focused on issues of environmental racism and the ways in which various working-class and excluded communities were being denied their basic environmental rights. Gradually, these disparate groups started to consolidate around a common cause and agreed a broad set of political and moral principles. It is these principles of environmental rights and responsibility which provide the context within which the contemporary Environmental Justice Movement now operates.

Key reading: Bullard, R.D. (2001) 'Anatomy of environmental racism and the Environmental Justice Movement', in J.S. Dryzek and D. Schlosberg (eds) *Debating the Earth: The Environmental Politics Reader*: 471–492; Bullard, R.D. (1990) *Dumping in Dixie: Race, Class and Environmental Quality*.

sanctioned visions of sustainable citizenship which appear to fit so nicely with the idea of locally based sustainable communities. I want to claim that a more radical (and perhaps more literal) interpretation of sustainable citizenship challenges many of the geographical and political assumptions associated with existing models of citizenship.

Unpacking sustainable citizenship

Given our discussion of the different modes and forms of citizenship in the previous section, I want to begin by asking a simple question – what is a sustainable citizen? I came across one interesting interpretation of sustainable citizenship in a new school curriculum for global citizenship and sustainable development. It suggested that sustainable citizenship emphasized:

> *Interdependence* – understanding how people, the environment and the economy are inextricably linked at all levels from the local to the global; *Stewardship* – recognising the importance of taking individual responsibility and action to make the world a better place; *Diversity* – understanding, respecting and valuing both human diversity – cultural, social and economic – and bio-diversity.
>
> (Welsh Assembly Government 2002: 9,
> quoted in Bullen and Whitehead 2005)

Within this quote we can begin to discern the key character traits of the sustainable citizen. First, the idea of *interdependence* suggests a citizen who is keenly aware of the ways in which social, economic and environmental processes connect and how the activities of local and national communities can impact on other more distant peoples (Bullen and Whitehead 2005) (see Figure 8.3). The notion of *stewardship* appears to emphasize the importance of taking an active role in ensuring sustainability in the context of everyday life and community relations. Finally, the focus in this quote on *diversity* illustrates that a sustainable citizen is someone who recognizes social and environmental needs and attempts to secure social and ecological justice.

Drawing on these principles of *interdependence, stewardship* and *diversity* as illustrative examples of what being a sustainable citizen may involve, it is clear that the ideal of sustainable citizenship challenges many of the established norms associated with state-centric modes of citizenly rights and responsibilities. Elsewhere, I have argued that the idea of sustainable citizenship suggests the *stretching* or expanding of the traditional rights and responsibilities associated with the national citizen (Bullen and Whitehead 2005) (see Figure 8.4). At one level this stretching process

Figure 8.3 Environmental knowledge and the sustainable citizen

can be thought of geographically. As we have seen throughout this book, the principles and philosophies of sustainability prioritize a concern for social, economic and environmental conditions throughout the world. In this context, the sustainable citizen it would seem is a citizen who is not just concerned with the welfare of his/her immediate local community or fellow nationals, but takes an active interest in the welfare of environments and people who live in distant parts of the world. This concern for distant others, which appears to be such an important part of a sustainable citizen's identity, resonates strongly with emerging forms of post-national or cosmopolitan citizenship (Desforges 2004; Linklater 1998; Turner 2000). These new models of citizenship suggest that systems of rights and responsibilities should not be constructed around exclusionary national boundaries, but should take a much more inclusive and planetary style. At a second level the stretching process encapsulated within the principles of sustainable citizenship requires a reconsideration of the relationship between citizenship and time. Conventionally, the rights and responsibilities associated with citizenship have been ascribed to living subjects – that is the current generation of citizens. Sustainable development's emphases on the needs of future generations suggests that when calculating the rights of citizenship, we should also consider the rights of unborn generations (see Barry 1999; Bullen and Whitehead 2005: 502). A final dimension of the stretching which is implicit within the ideals of sustainable citizenship

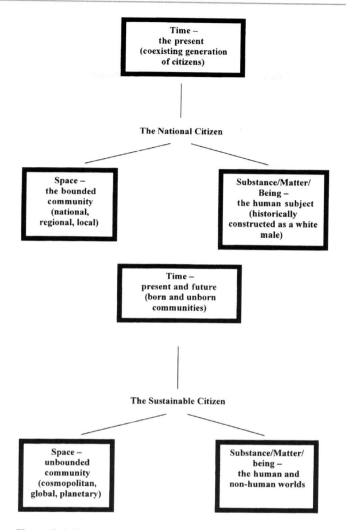

Figure 8.4 The national and sustainable citizen

relates to the role of the environment and the non-human world within prevailing modes of rights and responsibilities (see Bullen and Whitehead 2005: 504 and 507). Although existing forms of citizenship are constructed almost exclusively around the rights of human subjects and the responsibilities of humans towards other humans, the notion of sustainability suggests a greater awareness of the rights and responsibilities which flow between the human and non-human world (ibid.). While it is perhaps difficult to image a system of shared socio-ecological citizenship (see Dobson 2003: 13), it is clear that the interdependence of the human and non-human

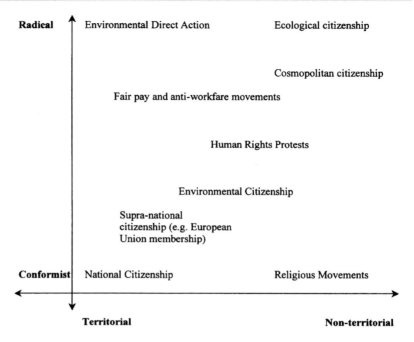

Figure 8.5 A grid of citizenship types

worlds does suggest the existence of a broad social and ecological community of being (Bullen and Whitehead 2005: 504). As a form of stretched, socio-environmental citizenship, it is clear that sustainable citizenship is linked to a range of other citizenship types. These citizenship types are more-or-less territorial and more-or-less radical (see Figure 8.5). While sustainable citizenship has clearly drawn inspiration from many of these different types of citizenships, I want to claim that its radical potential reaches far beyond any pre-existing models of citizenship.

Exploring the geographies of the sustainable citizen

In this final section I want to explore two examples of sustainable citizenship in practice. The first considers sustainable citizenship in a local context through a more detailed consideration of the activities associated with Local Agenda 21. The second case study considers the emergence of sustainable citizenship education in Welsh schools. In both instances we will see the effects which prevailing geographical assumptions and state–civil society relations have on the types of sustainable citizenship which are currently being promoted.

Local citizenship and Agenda 21

In many ways the Local Agenda 21 (LA21) programme I discussed earlier in this chapter is a meeting place for debates concerning sustainable citizenship and localities. As previously stated, LA21 was conceived as the local operational arm of Agenda 21 – the vision of sustainable development established at the Rio Earth Summit (UNCED). Essentially LA21 has involved the transfer of key aspects of sustainable development policy from international organizations and national states to local governments. While different local governments have obviously interpreted LA21 differently – particularly in the context of their own local priorities (see Chapters 2–5) – generally LA21 has involved local authorities creating local plans for sustainable development. Within these local sustainable development plans, local authorities set out how they intend to address issues of sustainability within their local areas and incorporate sustainable development considerations within their policy making.

Despite its widespread proliferation, initially there was very little precise guidance on the direction and form which LA21 schemes should take. It was in this context that the International Council for Local Environmental Initiatives (hereafter ICLEI) established a declaration and guidelines, giving more precise details of the roles and functions of LA21 programmes.[4] Within this declaration the ICLEI established the broad policy parameters within which LA21 initiatives should operate (including transport, planning, water supply, waste disposal, etc.) and stressed the importance of a community-centred and multi-agency approach within LA21 programmes. The Aalborg Charter (1994) – which seeks to broadly apply ICLEI principles to European urban municipal authorities – provides further insights in to the principles of LA21. Within the Aalborg Charter we see precisely why local authority-based forms of sustainability are believed to be so important. The Charter states:

> [We] are convinced that sustainable human life on this globe cannot be achieved without sustainable local communities. Local government is close to where environmental problems are perceived and closest to the citizens and shares responsibility with governments at all levels for the well-being of humankind and nature.
>
> (Aalborg Charter 1994)[5]

In this context, the Aalborg Charter reasserts the importance of producing a long-term sustainable development plan for local areas. The stages of plan formation established by the charter are illustrative of the goals of the LA21 process. The Aalborg Charter consequently suggests that there are eight key stages within the development of a local plan for sustainable development:

1   Identifying pre-existing plans and local institutions in the planning area;
2   Identification of key social, economic and environmental problems facing a local area through public consultation;
3   Prioritization of key local issues identified in the consultation process;
4   Production of a vision of local sustainable development again through widespread consultation;
5   Assessment and analysis of alternative visions for the creation of a local sustainable community;
6   Formation of a long-term plan for local sustainable development (which should include measurable goals, milestones and targets);
7   Development of a programme of implementation for the plan (which should include issues of funding and staffing);
8   Creation of a formal system of procedures designed to monitor the implementation of the plan.

(adapted from Aalborg Charter 1994)

What we can deduce from these broad objectives of LA21 plan formation is threefold. First, it is clear that LA21 is something which is at least initially to be led and guided by local government. Second, LA21 is to be based upon the creation of multi-sector partnerships, within which members of local public, private and voluntary sectors have a role in contributing to the formation of a vision of local sustainable development and of then implementing that vision. In this context, it is clear that LA21 should play an important role in effectively co-ordinating the activities of disparate organizations in an attempt to create a more sustainable local community. Third, and finally, the local modes of citizenship associated with LA21 schemes are very much centred around the state and the notion of the citizen as a user of state services.

Having established what LA21 planning involves in principle, it is important to consider its practical implementation and implications. Jörby and Lindström's analysis of local sustainability initiatives in Sweden is one of the most extensive reviews of LA21 schemes currently available (see Jörby 2002). This study was based upon a five-year study of LA21 programmes in four Swedish municipalities. Jörby and Lindström's analysis of LA21 schemes in Sweden is significant not only because of its temporal and spatial scope, but also because the Swedish constitution ensures that Sweden has one of the most powerful and influential sets of local government authorities in Europe. Focusing on four local government areas in southern Sweden, Jörby and Lindström's study revealed that LA21 has created new opportunities for sustainable development planning and sustainable citizenship. At one level it appears that the emphasis on integrated social, economic and environmental planning involved in the LA21 process has created an impetus for different local government departments to co-ordinate their activities more effectively in Sweden. Jörby and Lindström's study also illustrates that LA21 planning has provided the context within which the environment is becoming a major factor in a range of previously unrelated policy areas, including economic development, housing and health care. Finally, the evidence gathered from Swedish municipalities also suggests that the formation and implementation of LA21 plans has created a situation whereby public consultation and participation in local decision making has been broadened and enhanced (Jörby 2002: 223). It appears that the new emphasis on sustainable development within Swedish local government has provided new opportunities for sustainable citizenship to emerge. The actual results of these new procedures for the delivery of sustainable development policies on the Swedish environment are discussed at length by Jörby (2002). It is claimed that new forms of inter-governmental co-ordination and public participation in local environmental policy making can be directly linked to increased household recycling rates, reduced air pollution levels, improvements in the quality of fresh water and marine ecosystems, and the formation of a range of new environmental protection areas (Jörby 2002)

As well as revealing the positive impacts which LA21 has made to the operation of local government in Sweden, Jörby and Lindström's study has also exposed some of the limitations of LA21. Significantly in the context of this chapter, the limitations identified by Jörby and Lindström are directly related to issues of locality and citizenship. First, Jörby and Lindström emphasize that LA21 planning in Sweden is very state-centric and depends heavily on local government support (Jörby 2002: 233) While at one level this may be expected, it is clear that the strong emphasis on the role of local government in the LA21 process in Sweden has had a big impact on the forms of citizenship it appears to be generating. As a largely top-down initiative, LA21 in Sweden tends only to engage with local people in an intermittent and consultative way. This situation appears to have resulted in a position whereby more active, and even radical, modes of citizenly engagement in local sustainability have been severely limited in many Swedish municipalities. At one level, this lack of sustained citizenly engagement in the LA21 process may appear to be a necessary consequence of the programme being state-led. In turn this issue is offset by the ability of local government districts to use their financial, institutional and planning powers to promote sustainable development in ways that individuals never could. This said it appears unlikely that LA21 projects will ever be sustainable without eventually being transmitted into the local community sphere. Given the fact that LA21 programmes remain under-staffed and underfunded in Sweden, it appears even more urgent that local municipalities find new ways of enabling local citizen to take control of sustainable development initiatives (see Jörby 2002: 233).

In addition to failing to promote more active and sustainable modes of civic engagement in local politics, Swedish LA21 schemes also appear to be having problems with their *local* constitution. As we have heard consistently throughout this chapter, the great value of LA21 schemes is that they focus on localities, or those places and spaces which people feel their greatest sense of attachment to and control over. However, in Sweden, as with many other countries, the 'local' in LA21 programmes has been taken largely for granted. The local has routinely been interpreted as the local government area or boundary within which a particular LA21 plan operates. As we noted in the first section of this chapter, however, people's attachment to place and localities cannot be mapped onto arbitrary political boundaries. Indeed many people feel no particular sense of attachment or responsibility towards their local administrative area. People are attached to their neighbourhoods, micro-communities and other more organic places. The prioritization of local government boundaries in Swedish (and other countries') LA21 programmes has consequently had two negative implications. First, the sense of the 'local' promoted in these schemes has not really engaged the imagination of local people or encouraged them to become involved in sustainable development initiatives. Second, the narrow emphasis on local political areas appears – at least in the case of Sweden – to have prevented any meaningful sustainable development planning across local boundaries. With many different types of political, cultural and ecological communities crossing local political boundaries, this issue appears to present a significant barrier to effective sustainable development in the future.

What the case of LA21 planning in Sweden emphasizes very well is that there is nothing intrinsically good about the local when it comes to attempting to generate a more sustainable society. As is the case with the region, localities do not pre-exist their construction and in this context any discussions of the link between sustainability and the local should start with the question 'which local'? In the case of Sweden it appears that a very rigid, formally defined notion of local space is being deployed within LA21 plan formation. While such depictions of the local have many administrative and legal benefits, it is also clear that they can inhibit the ability of sustainable development to embed itself in local communities and in the actions of local citizens.

Reconstituting national citizenship: education for global citizenship and sustainable development in Wales[6]

As a lecturer I have always been interested in the ways in which people teach about issues concerning sustainability and citizenship. I suppose my interest stems from a concern I have that while notions of sustainability and citizenship can be portrayed as relatively simply categories within education, this assumption actually belies their contested political constitutions (see Whitehead 2003d). Given this concern I became particularly interested in a Welsh government initiative to promote education for sustainable citizenship throughout Wales. At one level I was delighted that such an important issue would be addressed in the mainstream educational curriculum in Wales. At the same time, however, I was concerned about how notions of sustainable citizenship would actually be taught in schools and what types of sustainable citizenship would be promoted.

The promotion of education for sustainable citizenship in Wales must be placed in the context of devolution. In 1999 the British government devolved certain key powers from London to Cardiff and formed the National Assembly for Wales. The National Assembly for Wales is an elected assembly of Welsh politicians who post-1999 have responsibility for key areas of public policy in Wales. There are two important things to note about the Government of Wales Act (1998) that laid the foundations for the National Assembly for Wales. First, it made the delivery of sustainable development a statutory obligation of the Welsh Assembly. Second, one of the key areas of policy decision making devolved from London to Cardiff was education. It is unquestionably the commitment of the Welsh Assembly to sustainable development and its power to shape the Welsh educational landscape that has given rise to a new focus on education for sustainable citizenship in Wales (for a more detailed analysis of this process see Bullen and Whitehead 2005). In a sense, education has provided a context within which the National Assembly for Wales has been able to promote sustainable development at a very local and individual level, while also raising awareness of sustainability among the next generation of young citizens.

Education for sustainable citizenship in Wales really began in 2002 when the Qualifications, Curriculum and Assessment Authority for Wales produced a revised curriculum. The curriculum was entitled *Education for Sustainable Development and Global Citizenship* (Welsh Assembly Government 2002). To put this new curriculum in context, it now means that young children going to school in Wales will not only be learning their times-tables, British history and the chemical composition of water, but also how to be more sustainable citizens (Bullen and Whitehead 2005: 501). These new modes of citizenship education focus upon explaining to young people their place in a much wider global community and exploring their social, economic and environmental relations with the wider world.

Despite appearing to offer a welcome and progressive alternative to the promotion of national modes of citizenship in schools (particularly through a focus on nationalist histories and the promotion of particular nationally inscribed civic values), closer inspection of what is occurring in Wales reveals the contested politics of sustainable citizenship. One contradiction which is particularly evident in Wales' education for sustainable citizenship scheme, is that while it is ostensibly global in its outlook, it is intimately tied to the creation of a new Welsh, national citizen (see Bullen and Whitehead 2005). Consequently, Welsh education officials involved in the new curriculum consistently refer to the ways in which education for sustainable citizenship is contributing to the production of a new Welsh person, who is internationalist in outlook, but, perhaps more importantly, is very different from their English counterparts (ibid.: 511). The use of the notion of sustainable citizenship to bolster Welsh national identity illustrates that despite its stretched spatial outlook, sustainable citizenship can still be used to recreate territorial division. A second tension which has started to emerge around Wales' education for sustainable citizenship scheme relates to what should actually be taught in the curriculum. While certain government officials are keen to create a standardized 'national' curriculum – perhaps addressing the erosion of the rainforest in Brazil, climate change, or agricultural reform in India – others are opposed to such a move. In this context, certain education officials and teachers argue that the creation of a highly structured curriculum for sustainable development education goes against the very essence of sustainability. They argue that what is needed are a series of practical and locally specific projects, which directly link the school curriculum with issues in the local community and more distant communities with which the school can interact (see Bullen and Whitehead 2005: 512). Related to this tension regarding the creation of a standardized curriculum, there are others involved in the scheme who are questioning whether sustainable citizenship is really something which should be taught in the classroom in the first place. To some, teaching sustainable citizenship in the space of the classroom tends to suggest that it is a formal subject which is simply learned and executed in schools. Many argue that by linking sustainable citizenship to school education, it is becoming synonymous with traditional modes of civic education concerning public morals and etiquette (see Bullen and Whitehead 2005: 512).

In light of these criticisms, and reservations, it is clear to see that sustainable citizenship is not simply an ideal goal which can be unproblematically achieved. What a sustainable citizen is and how a sustainable citizenry should be formed are matters of political and ideological debate.

Crucially this case study reveals that the emergence of sustainable citizenship is intimately tied to the territories (i.e. Wales) and spaces (i.e. the classroom) through which it is pursued.

Summary

In this chapter we have considered some of the connections which are being constructed between sustainability and the local. In particular we have considered why the local is seen to be such an important geographical scale in and through which to foster a more sustainable society. By focusing on the LA21 process we have seen that the promotion of local forms of sustainability has been tied to the creation of a more sustainable citizenry of engaged and active communities. Crucially, in this context, we have seen how the idea of a sustainable citizen challenges and stretches the conventional spatial and temporal horizons associated with citizenship. While offering hope for a more radical and dynamic mode of transnational and trans-temporal citizenship, we have also seen that as sustainable citizenship is currently being constructed, many of its most progressive and innovative dimensions are being suppressed.

In the final section of this chapter we considered the cases of LA21 programmes in Sweden and education for sustainable citizenship in Wales. In both instances we discovered that the role of the state was having a significant impact on the models of sustainable citizenship which were beginning to emerge. We also found that the particular geographical context within which both programmes were operating (local government and national space respectively), were constraining the trans-local and supra-national ideals associated with sustainable citizenship. What both case studies clearly show is that in discussions of local sustainable development and sustainable citizenship, neither the local nor the citizen are pre-given entities. In this context, when we consider local sustainable development we must ask what is the local in this form of development, and when we discuss sustainable citizenship, we must also consider what form of citizenship we are envisioning for the sustainable society. These are important questions, but they appear to be questions which geographers are particularly well placed to analyse.

Suggested reading

While much has been written on the 'local' in geography, I would recommend as an accessible introduction to geographical approaches to localities: Massey, D. (1994) *Space, Place and Gender*. In terms of discussions of citizenship, for a good overview of and introduction to the politics of citizenship I would suggest reading selected chapters of Isin, E.F. and Turner, B.S. (eds) (2002) *Handbook of Citizenship Studies*. For an excellent and engaging explanation of the links between geography and citizenship see: Brown, M. (1997) *RePlacing Citizenship: AIDS Activism and Radical Democracy*. For a very detailed analysis of different modes and philosophies of citizenship and a discussion of the links between citizenship and the environmental movement consult: Dobson, A. (2003) *Citizenship and the Environment*. For a more specific analysis of sustainable modes of citizenship see: Bullen, A. and Whitehead, M. (2005) 'Negotiating the networks of space, time and substance: a geographical perspective on the sustainable citizen', *Citizenship Studies*, 9: 499–516.

Suggested websites

For more information on the philosophies and activities of the Environmental Justice Movement go to the Environmental Justice Fund website at: http://www.ejfund.org/ejmove/ejmove.htm#Anchor-Key-21683 (accessed on 13 February 2003).

For a detailed description of the principles and programmes of the International Council for Local Environmental Initiatives see: http://www.iclei.org/projserv.htm (accessed 17 November 2005).

9 CONCLUSION
Reflections on 'actually existing sustainabilities'[1]

Those of you with good memories will recall that I began this book by recounting my encounter with Yann Arthus-Bertrand's now famous Earth from the Air exhibition. As you will remember, this exhibition was designed to keep a photographic record of the planet Earth at the start of the new millennium. I talked about the Earth from the Air in the first chapter as an excellent example of the geographies of sustainabilities I have subsequently explored throughout this volume. What interests me about this exhibition – having now completed this book – is what an equivalent photographic exhibition of the Earth would look like in the year 3000? Perhaps the more pessimistic among us would argue that this question is pointless given that humankind will probably have run itself out of existence by this time. But to me this question is what this book has essentially been all about – a consideration of how contemporary strategies to build sustainable communities within different geographical locations throughout the world can contribute in different ways to the formation of a more sustainable global society and ecosystem. In considering this issue, we have travelled far and wide, from the Barents Sea in the Arctic, to equatorial Kenya; from the megalopolis that is Mexico City, to the sparsely populated Dyfi Valley in Mid Wales; from corporate boardrooms in the USA, to family farms in the Punjab.

In exploring these different places and geographical spaces we have considered both the *geography of sustainability* (how notions of sustainability vary from place to place) and the *new geographies* which are being produced in the light of sustainable development considerations (that is how concerns over sustainability are reshaping forms of geographical organization from cities to regions). In relation to the geographies of sustainability, we have seen the great variations which exist both within and between the MEDCs, LEDCs and post-socialist states when it comes to interpreting and implementing programmes for sustainable development. Geographical variations within the political and social interpretation of sustainability are not necessarily a bad thing. Different interpretations of sustainable development, for example, often reflect attempts to reshape notions of sustainability to fit with pre-existing socio-cultural and scientific traditions. What geographical variations in sustainability do represent, however, is a challenge to the overly historicized accounts of sustainable development which prevail in much of the non-geographical literature. Within these accounts, sustainability is reduced to the historical emergence of the singular concept of sustainable development, which was first consolidated by the World Commission on Environment and Development and has subsequently been dispersed throughout the world through global UN conferences and agreements. While global negotiations around the idea of sustainable development have clearly been vital in constituting the forms of sustainability we have been exploring, *actually existing sustainabilities* do not simply represent local manifestations of UN-sanctioned models of sustainable development – nor should they. While for some this realization may be a cause for concern, as they see an ideal type of sustainable development being diluted and perhaps even misinterpreted within different geographical contexts, to me, recognizing the diverse geographies of sustainability is a cause for celebration. It is a cause for celebration because it reflects that, far from being just a post-war, global political compromise, the desire to be sustainable actually has a much longer place in human history and related socio-cultural traditions. The question is whether globally agreed action for sustainable development (with all of its flaws) can continue to provide the stimulus for geographically meaningful and 'sustainable' sustainabilities to emerge.

In addition to these geographies of sustainability we have also considered how the principles of being sustainable are changing the geographical structures of our existence. The new geographies of sustainability can be seen in the form of sustainable cities and regions and transitional planning zones like that which we explored in the Barents Sea. These new geographies are, however, also finding expression in relation to changing geographical perceptions of the world. These shifting geographical imaginations can be clearly seen within the stretched senses of socio-ecological responsibility being promoted within emerging brands of sustainable citizenship, and the growing sense of global connectedness which is associated with the discourses of the sustainable citizen. While it is relatively easy to chart the emergence of these new discourses of sustainability, assessing the actual impacts of these new ways of apprehending the world is more difficult to achieve. At one level this difficulty is based on the problems of establishing absolute measures or indicators of sustainability. Given the cultural and geographical specificity of the sustainabilities described in this volume, the value of any such absolute measure is of course open to question. At another level, however, even when sustainability is measurable, it is difficult to ascertain whether improvements in absolute levels of sustainability are actually the product of policies specially designed to achieve sustainable development, or merely the inadvertent outcome of shifting political and economic processes. It appears that at present the new geographies of sustainability described in this volume are at best alleviating the unsustainability of existing socioeconomic practices. Much more, however, needs to be done if reductions in the levels of unsustainable practice are to be replaced by an actual increase in the aggregate sustainability of our global society and ecosystem.

Although this volume has described numerous projects and programmes for sustainability, there are worrying trends which appear to be threatening our collective ability to construct a more sustainable future. The escalation of resource wars over increasingly scarce environmental assets, the spread of HIV/AIDS throughout the world and the escalating incidence of extreme weather events like hurricane Katrina in the USA are all actively challenging sustainable forms of development. In the context of these emerging pressures, it appears that our ability to balance social, economic and environmental need is going to become more, not less, difficult in the future. Having said this, however, the appalling suffering associated with such events only serves to remind us of the importance of trying to build more sustainable futures and of becoming continually aware of how social, economic and environmental processes connect within different geographical communities and spaces.

NOTES

Chapter 1

1 While much of this early work tended to focus upon environmentalism, not sustainability per se, it is now clear that it was addressing the debates which were surrounding the embryonic concept of sustainability at the time.

2 The World Fact Book (2004) http://www.cia.gov/cia/publications/factbook/geos/us.html#Econ (accessed on 1 July 2004). *Guardian* (2003) 'Nestlé u-Turn on Ethiopian Debt' 4 January.

3 I do, however, recognize that it was a World Council of Churches conference (on Science and Technology for Human Development) in 1974 where the term 'sustainable society' was first used at a major international meeting (see Dresner 2002: 29).

4 One of the key scientific groups responsible for illustrating the links between environmental change and anthropogenic activities was the International Biological Programme (IBP). The IBP was established in 1964 by the International Union of Biological Sciences (IUBS) and the International Council of Scientific Unions (ICSU) (see Adams 1990b: 30–31).

5 While few comprehensive analyses of the Johannesburg summit are yet available, the United Nations has published a useful summary document of what was achieved at the summit (see United Nations 2003a).

Chapter 2

1 See http://www.thebodyshop.com/web/tbsgl/about.jsp

2 It is unclear what affect l'Oréal's (the large cosmetics manufacturer) acquisition of the Body Shop in 2006 will have on the retailer's established image as a socio-environmentally sustainable company.

3 The UK's first environmental strategy did, however, support the adoption of a *precautionary principle* in British environmental policy which had been absent during the 1970s and 1980s (HMSO 1990: 136).

4 This letter was sent to George W. Bush by representatives from thirty-one political groups in the USA and also urged him not to sign up to international environmental agreements being discussed at the summit.

5 The Pacific Legal Foundation is only one of a series of foundations campaigning against the restrictions imposed by government legislation on

landowners' use of environmental resources; see also for example the Mountain State Legal Foundation at: http://www.mountainstateslegal.org/
6   More information about the role and activities of the Pacific Legal Foundation can be found on their website at: http://www.pacificlegal.org
7   Joe Klein argues that although at times Clinton appeared to support the basic principles of the Kyoto protocol, in reality this was more of an attempt to build a broad international coalition of support for his foreign policy goals, than a signal that he would ever sign the agreement.

Chapter 3

1   See: http://www.unhabitat.org/programmes/sustainablecities/katowice.asp
2   For more on the projects associated with the SKAP visit: http://www. unhabitat.org/programmes/sustainablecities/katowice.asp.
3   However, some have argued that the initial isolation of the Soviet Union from the United Nations sustainable development programme has been overstated because of the continuing long-term input the USSR was able to have through the General Assembly of the UN (see Dresner 2002).
4   For more information about the Russian Federation's policies and laws for sustainable development visit the United Nations Department of Social and Economic Affairs website at: http://www.un.org/esa/agenda21/natlinfo/ countr/russia/

Chapter 4

1   For more information on global poverty visit the United Nations Statistics Division at: http://unstats.un.org/unsd/methods/poverty/default.htm (accessed 14 November 2005).
2   See: http://unstats.un.org/unsd/methods/poverty/default.htm (accessed 14 November 2005).
3   These and others statistics on global poverty are available through the United Nations Statistics Division at: http://millenniumindicators.un.org/ unsd/
4   See: http://millenniumindicators.un.org/unsd/
5   This figure is based on information taken from the Green Belt Movement's home page at: http://www.greenbeltmovement.org/
6   For more information on these projects go to: http://www.greenbelt movement.org/
7   See: http://www.greenbeltmovement.org/
8   See: http://www.greenbeltmovement.org/

Chapter 5

1   I am indebted to Jamie Peck for this phrase. He used it during a plenary session at the Annual Conference of the Association of American Geographers, Pittsburgh, 2001, and I have been fascinated with it ever since.

2 These figures are taken from the BBC's climate change page at: http://www.bbc.co.uk/climate/

3 Following his untimely death on 20 June 2005, *The Independent* newspaper (UK) ran an excellent obituary through which you can find out more about Charles David Keeling's life and work (*The Independent*, 27 June 2005: 32).

4 For more on the United Nations Convention on Climate Change go to: http://www.bbc.co.uk/climate/ (accessed 14 November 2005).

5 These figures are taken from the BBC Climate Change website at: http://www.bbc.co.uk/climate/ (accessed 14 November 2005).

Chapter 7

1 For more on health research in Stoke-on-Trent visit Stoke-on-line at: http://www.stoke.gov.uk/navigation/category.jsp?categoryID=370638 (accessed 16 November 2005).

2 For more information on social and environmental conditions in Mexico City see the United Nations-sponsored web page: http://www.un.org/cyber schoolbus/habitat/profiles/mexico.asp

3 Ibid.

Chaper 8

1 For an excellent and concise review of the issues relating to geographical theories of place and the local see Duncan, J. (2002) 'Place', in R. Johnston *et al.* (eds) *The Dictionary of Human Geography* (Oxford: Blackwell): 582–584. See also Agnew, J. and Duncan, J. (eds) (1990) *The Power of Place* (Boston, MA: Unwin Hyman).

2 For more information on Local Agenda 21 visit: http://www.crossroad.to/text/articles/la21_198.html (from where this quote is taken) (accessed 17 November 2005).

3 For a recent example of work considering the geographies of citizenship see the special issue of *Citizenship Studies*, 9(5) 2005 edited by L. Desforges, R. Jones and M. Woods.

4 The International Council for Local Environmental Initiatives is a UN-supported partnership of municipal authorities devoted to promoting sustainable development in local government. The ICLEI was established in 1990.

5 For more information about the Aalborg Charter and the ongoing process of sustainable urban development in Europe go to: http://www.global-vision.org/city/aalborg.html (accessed 28 June 2006).

6 For a more detailed analysis of the new curriculum for global citizenship and sustainable development in Welsh schools see Bullen and Whitehead 2005; this paper contains quotes from teachers and school children actually involved in the curriculum.

Chapter 9

1 This phrase is taken from a call for papers for the Annual Conference of the Association of American Geographers (17 August 2005) for the session 'Engaging critical spaces for sustainability'. The session organizers were: James Evans, David Gibbs, Rob Krueger and Terry Marsden.

BIBLIOGRAPHY

Adams, W.M. (2001) *Green Development: Environment and Sustainability in the Third World*, second edition, London: Routledge.

—— (1990a) 'Rationalization and conservation: ecology and the management of nature in the United Kingdom', *Transactions of the Institute of British Geographers*, 22: 277–291.

—— (1990b) *Green Development: Environment and Sustainability in the Third World*, first edition, London: Routledge.

Adams, W.M. and Hollis, G.E. (1989) *Hydrology and Sustainable Resource Development of a Sahelian Floodplain Wetland*, London: Hedejia-Nguru Wetlands Conservation Project.

Agnew, J. (1987) *Place and Politics: The Geographical Mediation of State and Society*, Boston, MA: Allen and Unwin.

Agnew, J. and Duncan, J. (eds) (1990) *The Power of Place*, Boston, MA: Unwin Hyman.

Agnew, J., Livingstone, D.N. and Rogers, A. (eds) (1996) *Human Geography: An Essential Anthology*, Oxford: Blackwell: 365–512.

Agyeman, J., Bullard, R.D. and Evans, B. (2003) *Just Sustainabilities: Development in an Unequal World*, London: Earthscan.

Allen, J., Massey, D. and Cochrane, A. (with Charlesworth, J., Court, G., Henry, N. and Sarre, P.) (1998) *Rethinking the Region*, London: Routledge.

Amin, A. (2002) 'Spatialities of globalization', *Environment and Planning A*, 34: 385–399.

Amin, A. and Thrift, N. (2002) *Cities – Reimagining the Urban*, Cambridge: Polity Press.

Amis, M. (2002) *Koba the Dread*, London: Jonathan Cape.

Andrusz, G., Harloe, M. and Szelenyi, I. (eds) (1996) *Cities After Socialism: Urban and Regional Change and Conflict in Post-Socialist Societies*, Oxford: Blackwell.

Bahro, R. (1986) *Building the Green Movement*, London: Heretic Books.

—— (1984) *From Red to Green*, London: Verso.

Baker, S. (2000) 'Between the devil and the deep blue sea: international obligations, eastern enlargement and the promotion of sustainable development in the European Union', *Journal of Environmental Policy and Planning*, 2: 149–166.

Baker, S. and Jehlicka, P. (eds) (1998) *Dilemmas of Transition: The Environment, Democracy and Economic Reform in East and Central Europe*, London: Frank Cass.

Baker, S., Richardson, D., Young S. and Kousis, M. (eds) (1997) *Sustainable Development: Theory, Policy and Practice within the EU*, London: Routledge.

Barry, B. (1999) 'Sustainability and intergenerational justice', in A. Dobson (ed.) *Fairness and Futurity: Essays on Environmental Sustainability and Social Justice*, Oxford: Oxford University Press.

Barry, J. (2001) 'Rudolf Bahro 1935–97', in J.A. Palmer (ed.) *Fifty Key Thinkers on the Environment*, London: Routledge: 269–273.

Barry, J. and Paterson, M. (2003) 'The British state and the environment: New Labour's ecological modernisation strategy', *International Journal of Environment and Sustainable Development*, 4: 237–249.

Bartelmus, P. (1986) *Environment and Development*, London: Allen and Unwin.

Becker, E. and Jahn, T. (eds) (1999) *Sustainability and the Social Sciences: A Cross-Disciplinary Approach to Integrating Environmental Considerations in Theoretical Reorientation*, London: Zed Books.

Berry, B. (1990) 'Urbanization', in B. Turner (ed.) *The Earth as Transformed by Human Action: Global and Regional Changes in the Biosphere over the Past 300 Years*, Cambridge: Cambridge University Press.

Bialasiewicz, L. (2002) 'The re-birth of Upper Silesia', *Regional and Federal Studies*, 12: 111–132.

Blaikie, P. (1985) *The Political Economy of Social Erosion in Developing Countries,* London: Longman.

Blaikie, P. and Brookfield, H. (1987) *Land Degradation and Society,* London: Methuen.

Blowers, A. and Pain, K. (1999) 'The unsustainable city', in S. Pile, C. Brook and G. Mooney (eds) *Unruly Cities?*, London: Routledge: 247–298.

Bookchin, M. (1991) *The Ecology of Freedom: The Emergence and Dissolution of Hierarchy*, New York: Black Rose Books.

—— (1986) *The Modern Crisis*, Philadelphia, PA: New Society Publishers.

—— (1980) *Towards an Ecological Society*, Montreal: Black Rose Books.

Boyer, C. (1997) *Dreaming the Rational City: The Myth of American City Planning*, London: MIT Press.

Bradshaw, M. and Stenning, A. (2004) 'Introduction: transformation and development', in M. Bradshaw and A. Stenning (eds) *East Central Europe and the Former Soviet Union: The Post-Socialist States*, London: Pearson, Prentice Hall: 1–32.

Bratton, S.P. (1983) 'The ecotheology of James Watt', *Environmental Ethics*, 5: 225–236.

Brenner, N. (2004) *New State Spaces: Urban Governance and the Rescaling of Statehood.* Oxford: Oxford University Press.

—— (1999) 'Beyond state-centrism? Space, territoriality, and geographical scale in globalization studies', *Theory and Society*, 28: 39–78.

Bridge, G. and Wood, A. (2005) 'Geographies of knowledge, practices of globalization: learning from the oil exploration and production industry', *Area*, 27: 199–208.

Brown, M. (1997) *RePlacing Citizenship: AIDS Activism and Radical Democracy*, New York: Guilford.

Brunstad, B., Manus, E., Swanson, P., Hønnelland, G. and Øverland, I. (2004) *Big Oil Playground, Russian Bear Preserve or European Periphery? The Russian Barents Sea Region towards 2015*, Delft: Eburon.

Bryson, B. (2004) *A Short History of Nearly Everything*, London: Black Swan.

Bulkeley, H. (2005) 'Reconfiguring environmental governance: towards a politics of scales and networks', *Political Geography*, 24: 875–902.

Bulkeley, H. and Betsill, M.M. (2005) 'Rethinking sustainable cities: multi-level governance and the "urban" politics of climate change', *Environmental Politics*, 14: 42–63.

Bullard, R.D. (2001) 'Anatomy of environmental racism and the Environmental Justice Movement', in J.S. Dryzek and D. Schlosberg (eds) *Debating the Earth: The Environmental Politics Reader*, Oxford: Oxford University Press: 471–492.

—— (1990) *Dumping in Dixie: Race, Class and Environmental Quality*, Boulder, CO: Westview Press.

Bullen, A. and Whitehead, M. (2005) 'Negotiating the networks of space, time and substance: a geographical perspective on the sustainable citizen', *Citizenship Studies*, 9: 499–516.

Burgess, J. (2005) 'Environmental knowledges and environmentalism', in P. Cloke, P. Crang and M. Goodwin (eds) *Introducing Human Geographies*, London: Arnold: 298–310.

—— (1990) 'The production and consumption of environmental Meanings in the mass media: a research agenda for the 1990s', *Transactions of the Institute of British Geographers*, 15: 139–161.

Carter, N. (2001) *The Politics of the Environment: Ideas, Activism and Policy*, Cambridge: Cambridge University Press.

Castells, M. (2000) *The Rise of the Networked Society*, Oxford: Blackwell.

—— (1977) *The Urban Question: A Marxist Approach*, London: Edward Arnold.

Chatterjee, P. and Finger, M. (1994) *The Earth Brokers: Power, Politics and World Development*, London: Routledge.

Churchill, R.R. and Robin, R. (1992) *Marine Management in Disputed Areas: The Case of the Barents Sea*, London: Routledge.

Cliff, T. (1974) *State Capitalism in Russia*, London: Pluto Press.

Cloke, P., Crang, P. and Goodwin, M. (eds) (1999) *Introducing Human Geographies*, London: Arnold.

Coggins, G.C., and Nagel, D.K. (1990) 'Nothing beside remains: the legal legacy of James G. Watt's tenure as Secretary of the Interior on Federal Land Law and Policy', *Boston College Environmental Affairs Law Review*, 17: 473–550.

Collinge, C. (2005) 'The *difference* between society and space: nested scales and the returns of spatial fetishism', *Environmental and Planning D: Society and Space*, 23: 189–206.

Cosgrove, D. (2001) *Apollo's Eye: A Cartographic Genealogy of the Earth in the Western Imagination*, Baltimore, MD: Johns Hopkins University Press.

Cowell, R. (1997) 'Stretching the limits: environmental conservation, habitat creation and sustainable development', *Transactions of the Institute of British Geographers*, 22: 292–306.

Cresswell, T. (2006) *On the Move: Mobility in the Modern West*, New York: Routledge.

—— (2004) *Place: A Short Introduction*, Oxford: Blackwell.

—— (2001) *The Tramp in America*, London: Reaktion Books: 9–22.

—— (1996) *In Place/Out of Place*, London: University of Minnesota Press.

Dalby, S. (1996) 'Reading Rio, writing the world: the *New York Times* and the 'Earth Summit'', *Political Geography*, 15: 593–614.

Dasmann, R.F. (1985) 'Achieving the sustainable use of species and ecosystems', *Landscape Planning*, 12: 211–219.

DEFRA (2004) *Securing the Future – UK Government Sustainable Development Strategy*, London: DEFRA.

—— (2003) *The Environment in your Pocket 2003*, London: DEFRA.

Demeritt, D. (2001) 'The construction of global warming and the politics of science', *Annals of the Association of American Geographers*, 91: 307–337.

Desforges, L. (2004) 'The formation of global citizenship: international non-governmental organisation in Britain', *Political Geography*, 23: 549–569.

DETR (1999) *A Better Quality of Life: A Strategy for Sustainable Development in the UK*, London: DETR.

Dicken, P. (2002) 'Globalization', in R.J. Johnston, D. Gregory, G. Pratt and M. Watts (eds) *The Dictionary of Human Geography*, Oxford: Blackwell: 315–316.

Dobson, A. (2003) *Citizenship and the Environment*, Oxford: Oxford University Press.

—— (2000) 'Environmental citizenship', *Town and Country Planning*, January: 21.

—— (1995) *Green Political Thought*, London: Routledge.

—— (ed.) (1991) *The Green Reader*, London: Andre Deutsch.

Dodds, K. (2000) *Geopolitics in a Changing World*, Harlow: Prentice Hall.

DoE (1994) *Sustainable Development: The UK Strategy*, London: HMSO.

Doyle, T. and McEachern, D. (1998) *Environment and Politics*, London: Routledge.

Dresner, S. (2002) *The Principles of Sustainability*, London: Earthscan Publications.

Dryzek, J. and Schlosberg, D. (2001) *Debating the Earth: The Environmental Politics Reader*, Oxford: Oxford University Press.

Duncan, J. (2002) 'Place', in R. Johnston, D. Gregory, G. Pratt and M. Watts (eds) *The Dictionary of Human Geography*, Oxford: Blackwell: 582–584.

Duncan, J. and Savage, M. (1989) 'Space, scale and locality', *Antipode*, 21: 179–206.

Ecodyfi (2002) *Development and Action Plan*. Online. Available at: http://www.ecodyfi.org.uk/ecodyfireportspage.htm (accessed 6 April 2005).

Eden, S. (2005) 'Global and local environmental problems', in P. Cloke, P. Crang and M. Goodwin (eds) *Introducing Human Geographies*, London: Arnold: 271–284.

—— (2000) 'Environmental issues: sustainable progress?', *Progress in Human Geography*, 24: 111–118.

—— (1994) 'Using sustainable development: the business case', *Global Environmental Change*, 4: 160–176.

Ehrlich, P.R. and Ehrlich, A.H. (1996) *The Betrayal of Science and Reason*, Washington, DC: Island Press: 11–23.

Elliot, J. (1999) *An Introduction to Sustainable Development*, London: Routledge.

Environment Agency (2004) *Environmental Facts and Figures*, London: Environment Agency.

Environmental Protection Agency (2004) *Mexico City Diesel Retrofit Project*, Washington, DC: Environmental Protection Agency.

—— (1990) *The Environmental Protection Agency: A Retrospective*, EPA Press Release, 29 November.

Eskeland, G.S. (1992) 'Attacking air pollution in Mexico City', *Finance and Development*, 29: 28–30.

Fukuyama, F. (1993) *The End of History and the Last Man*, London: Penguin Books

Geddes, P. (1906) *A First Visit to the Outlook Tower*, Edinburgh: Geddes and Colleagues.

Gibbs, D. (2000) 'Ecological modernisation, regional economic development and regional development agencies', *Geoforum*, 31: 9–19.

Girard, L.F., Forte, B., Cerreta, M., De Toro, P. and Forte, F. (eds) (2003) *The Human Sustainable City: Challenges and Perspectives from the Habitat Agenda*, Aldershot: Ashgate.

Gore, A. (1992) *Earth in the Balance: Forging a New Common Purpose*, London: Earthscan.

Government Office for the West Midlands (2000) *Quality of Life, the Future Starts Here: A Sustainable Development Strategy for the West Midlands*, Birmingham: red fox media.

Graham, S. and Marvin, S. (2003) *Splintering Urbanism: Networked Infra-structures, Technological Mobilities and the Urban Condition*, London: Routledge.

Grainger, A. (1982) *Desertification: How People Make Deserts, How They Can Stop, and Why They Don't*, London: Earthscan.

Green Belt Movement (2005) *GBM Kenya: Environmental Conservation/Tree Planting*. Online. Available at http://gbmna.org/w.php?id=13 (accessed 28 June 2006).

Gregory, D. (2004) *The Colonial Present: Afghanistan, Palestine, Iraq*, Oxford: Blackwell.

Gregory, P.R. and Stuart, R.C. (1998) *Soviet and Post-Soviet Economic Structure and Performance*, sixth edition, New York: Harper Collins.

The Guardian (2002) 'Condemned to beating around the Bush', 24 August: 4.

The Guardian (2000a) 'Breathing space', *Guardian Unlimited Archive*. Online. Available at: http://www.guardian.co.uk/Archive/0,4273 (accessed 5 November 2005).

The Guardian (2000b) 'Protests erupt in violence', *Guardian Unlimited Archive*. Online. Available at: http://www.guardian.co.uk/Archive/0,4373 (accessed 5 November 2005).

Hacking, I. (1999) *The Social Construction of What?* Cambridge, MA: Harvard University Press.

Hajer, M. (1997) *The Politics of Environmental Discourse: Ecological Modernisation and the Policy Process*, Oxford: Clarendon Press.

Hajer, M. and Fisher, F. (1999) *Living with Nature: Environmental Politics as Cultural Discourse*, Oxford: Oxford University Press.

Hall, P. (2003) 'The sustainable city in an age of globalization', in L.F. Girard, B. Forte, M. Cerreta, P. De Toro and F. Forte (eds) *The Human Sustainable City: Challenges and Perspectives from the Habitat Agenda*, Aldershot: Ashgate: 55–70.

—— (1999) *Sustainable Cities or Town Cramming?* London: Entec.

—— (1996) *Cities of Tomorrow*, second edition, Oxford: Blackwell.

Hardin, G. (1968) 'The tragedy of the commons', *Science*, 162: 1243–1248.

Hardoy, J. and Satterthwaite, D. (1992) 'Towns and cities – empowering people', in J. Porritt (ed.) *Save the Earth*, London: Dorling Kindersley.

Hardt, M. and Negri, N. (2000) *Empire*, London: Harvard University Press.

Hardy, S. and Lloyd, G. (1994) 'An impossible dream? Sustainable regional economic and environmental development', *Regional Studies*, 28: 773–780.

Harrison, P. (1993) *The Third Revolution: Population, Environment and a Sustainable World*, London: Penguin Books.

Harvey, D. (2003a) *The New Imperialism*, Oxford: Oxford University Press.

—— (2003b) *Paris: City of Modernity*, London: Routledge.

—— (2000) *Spaces of Hope*. Edinburgh: Edinburgh University Press.

—— (1996) *Justice, Nature and the Geography of Difference*, Oxford: Blackwell.

—— (1989) *The Urban Experience*, Oxford: Blackwell.

—— (1985a) *Consciousness and the Urban Experience*, Oxford: Blackwell.

—— (1985b) *The Urbanisation of Capital*, Oxford: Basil Blackwell.

Haughton, G. and Counsell, D. (2004) 'Regions and sustainable development: regional planning matters', *Geographical Journal*, 170: 135–145.

Haughton, G. and Hunter, C. (1994) *Sustainable Cities,* London: Jessica Kingsley (with the Regional Studies Association).

Hayat, R.S. (2004) 'Poverty alleviation', in *Introducing Sustainable Development in 9½ Chapters*, London: Lead International (CD ROM).

Held, D., McGrew, A., Goldblatt, D. and Perraton, J. (1999) *Global Transformations – Politics, Economics and Culture*, Cambridge: Polity Press.

Henry, N. and Sarre, P. (1998) *Rethinking the Region*, London: Routledge.

Hetherington, K. (1997) *The Badlands of Modernity: Heterotopia and Social Ordering*, London: Routledge.

Hicks, B. (1996) *Environmental Politics in Poland: A Social Movement Between Regime and Opposition*, New York: Columbia University Press.

Hirst, P. and Thompson, G. (1996) *Globalization in Question*, Cambridge: Polity Press.

HMSO (1990) *This Common Inheritance: Britain's Environmental Strategy*, London: HMSO.

Holmberg, J. (1993) *Facing the Future: Beyond the Earth Summit,* London: Earthscan.

Huber, J. (2000) 'Towards industrial ecology: sustainable development as a concept of ecological modernization', *Journal of Environmental Policy and Planning*, 2: 269–285.

Huxley, A. (1994) *Brave New World*, London: Flamingo.

Ingold, T. (1993) 'Globes and spheres: the typology of environmentalism', in M. Milton (ed.) *Environmentalism – the View from Anthropology*, London: Routledge: 31–42.

International Monetary Fund (2003) *Status Report on Preparatory Activities and Way Forward for the Economic Recovery Strategy Paper (ers) for Kenya*, Washington, DC: International Monetary Fund.

International Union for the conservation of Nature and Natural Resources (1980) *The World Conservation Strategy*, Geneva: IUCN, UNEP and the WWF.

Isin, E.F. (2002) *Being Political*, Minneapolis: University of Minnesota Press.

Isin, E.F. and Turner, B.S. (eds) (2002) *Handbook of Citizenship Studies*, London: Sage.

IUCN (International Union for the Conservation of Nature and Natural Resources) (1980) *The World Conservation Strategy*, Geneva: IUCN, UNEP and the WWF.

Jehlicka, P. (2001) 'The new subversives: Czech environmentalists after 1989', in H. Flam (ed.) *Pink, Purple, Green: Women's, Religious, Environmental, and Gay/Lesbian Movements in Central Europe Today*, Boulder, CO: East European Monographs.

Jehlicka, P. and Cowell, R. (2003) 'Czech minerals policy in transformation: the search for legitimate policy approaches', *Environmental Sciences,* 1: 79–109.

Jehlicka, P. and Tickle, A. (2004) 'The environmental implications of eastern enlargement of the European Union: the end of progressive EU environmental policy?', *Environmental Politics,* 13: 77–95.

Jehlicka, P., Sarre, P. and Podoba, J. (2005) 'The Czech environmental movement's knowledge interests in the 1990s: compatibility of Western influences with pre-1989 perspectives', *Environmental Politics,* 14(1): 64–82.

Johnstone, C. and Whitehead, M. (eds) (2004) *New Horizons in British Urban Policy: Perspectives on New Labour's Urban Renaissance*, Aldershot: Ashgate.

Jones, M. (1998) 'Restructuring the local state: economic governance or social regulation?' *Political Geography,* 17: 959–988.

Jones, M. and MacLeod, G. (2004) 'Regional spaces, spaces of regionalism: territory, insurgent politics, and the English question', *Transactions of the Institute of British Geographers,* 29: 433–452.

Jones, M. and MacLeod, G. (1999) 'Towards a regional renaissance? Reconfiguring and rescaling England's economic governance', *Transactions of the Institute of British Geographers,* 24: 295–314.

Jörby, S.A. (2002) 'Local Agenda 21 in four Swedish municipalities: a toll towards sustainability?', *Journal of Environmental Planning and Management,* 45: 219–244.

Jordan, A., Wurzel, R.K.W. and Zito, A.R. (2005) 'The rise of "new" policy instruments in comparative perspective: has governance eclipsed government?', *Political Studies,* 53: 477–496.

Jordan, A., Wurzel, R.K.W. and Zito, A.R. (2003) 'New instruments of environmental governance: patterns and pathways of change', *Environmental Politics,* 12: 1–24.

Jubilee Debt Campaign (2002) *Drop the Debt Flyer*, London: Jubilee Debt Campaign.

Kabala, S. (1991) 'The environment and economics of Upper Silesia', *Report on Eastern Europe,* 2: 18–23.

Katowice Voivodship (1991) *Statistical Year Book of the Katowice Voivodship 1991*, Silesia: Katowice Voivodship.

Kearns, A. (1992) 'Active citizenship and urban governance', *Transactions of the Institute of British Geographers,* 17: 20–34.

Khanin, G. (1992) 'Economic growth in the 1980s', in M. Ellman and V. Kontorovich (eds) *The Disintegration of the Soviet System*, London: Routledge: 73–85.

Kirkby, J., O'Keefe, P. and Timberlake, L. (eds) (1995) *The Earthscan Reader in Sustainable Development*, London: Earthscan.

Klein, N. (2001) *No Logo*, London: Flamingo.

Kohr, L. (1957) *The Breakdown of Nations*, London: Routledge & Kegan Paul.

Kropotkin, P. (1912) *Fields, Factories, and Workshops: Or, Industry Combined with Agriculture and Brain Work with Manual Work*, London: Thomas Nelson & Sons.

Kuron, J. and Modzelewski, K. (1982) *Open Letter to the Party*, London: Bookmarks Publishing Co-operative.

Latour, B. (1993) *We Have Never Been Modern*, London: Harvester Wheatsheaf.

Lead (2004) *Introducing Sustainable Development in 9½ Chapters*, London: Lead International (CD ROM).

Lewis, J. (1985) 'The birth of EPA', *Environmental Protection Agency Journal Online*. Available at: http://www.epa.gov/history/topics/epa/15c.htm (accessed on 5 September 2006).

Lewycka, M. (2005) *A Short History of Tractors in Ukrainian*, London: Viking.

—— (1998) 'Cosmopolitan citizenship', *Citizenship Studies*, 2: 23–41.

Liverman, D.M. (1999) 'Geography and the global environment', *Annals of the Association of American Geographers*, 89: 107–120.

Lomborg, B. (2001) *The Skeptical Environmentalist: Measuring the Real State of the World*, Cambridge: Cambridge University Press.

Lovelock, J. (1979) *Gaia: A New Look at Life on Earth*, in J. Lovelock (1988) *Ages of Gaia*, Oxford: Oxford University Press.

Lowe, P. and Flynn, A. (1989) 'Environmental politics and policy in the 1980s', in J. Mohan (ed.) *The Political Geography of Contemporary Britain*, London: Macmillan.

Luccarelli, M. (1995) *Lewis Mumford and the Ecological Region: The Politics of Planning*, New York: Guilford Press.

Luke, T. (1999a) *A Rough Road out of Rio: The Right-wing Reaction in the United States Against Global Environmentalism* (mimeograph).

—— (1999b) 'Environmentality as green governmentality', in E. Darier (ed.) *Discourses of the Environment*, Oxford: Blackwell: 121–151.

Maathai, W. (2004) *The Green Belt Movement: Sharing the Approach and the Experience*, New York: Lantern Books.

McCormick, J. (1991) *British Politics and the Environment*, London: Earthscan.

McKibben, B. (2003) *The End of Nature: Humanity, Climate Change and the Natural World*, London: Bloomsbury.

MacLeod, G. (1998) 'In what sense a region? Place hybridity, symbolic shape, and institutional transformation in (post)-modern Scotland', *Political Geography*, 17: 833–863.

MacLeod, G. and Goodwin, M. (1999) 'Space, scale and state strategy: towards a rethinking of urban and regional governance', *Progress in Human Geography*, 23: 503–527.

McManus, P. (1996) 'Contested terrains: politics, stories and discourses of sustainability', *Environmental Politics*, 5: 48–71.

MacShane, D. (1981) *Solidarity: Poland's Independent Trade Union*, Nottingham: Spokesman.

Marsden, T., Banks, J., Renting, H. and van der Ploeg, J.D. (2001) 'The road towards Sustainable Rural Development: issues of theory, policy and research practice', *Journal of Environmental Policy and Planning*, 3: 75–85.

Marston, S. (2000) 'The social construction of scale', *Progress in Human Geography* 24: 219–242.

Marston, S. and Staeheli, L.A. (eds) (1994) 'Restructuring citizenship', *Environment and Planning* A, 26 (special issue).

Martinez-Alier, J. (2003) 'Urban sustainability and environmental conflict', in L.F. Girard, B. Forte, M. Cerreta, P. De Toro and F. Forte (eds) *The Human Sustainable City: Challenges and Perspectives from the Habitat Agenda*, Aldershot: Ashgate: 89–106.

Masanganise, P. and Swaminathan (2004) 'Food security and agriculture', in *Introducing Sustainable Development in 9½ Chapters*, London: Lead International (CD ROM).

Massey, D. (2004) 'Geographies of responsibility', *Geografiska Annaler Series B: Human Geography*, 86: 5–18.

—— (2000) 'Cities in the world', in D. Massey, J. Allen and S. Pile (eds) *City Worlds*, London: Routledge.

—— (1994) *Space, Place and Gender*, Cambridge: Polity Press.

—— (1993) 'Questions of locality', *Geography*, 78: 142–149.

May, J. (1996) 'Globalization and the politics of place: place and identity in an inner London neighbourhood', *Transactions of the Institute of British Geographers*, 21: 194–215.

Meadows, D.H., Meadows, D.L., Randers, J. and Behrens, W.W. (1972) *Limits to Growth: A Report for the Club of Rome's Project on the Predicament of Mankind*, New York: New York American Library.

Merchant, C. (2004) *Reinventing Eden: The Fate of Nature in Western Culture*, London: Routledge.

Mohan, J. (ed.) (1989) *The Political Geography of Contemporary Britain*, London: Macmillan.

Mol, A.P.J. (1999) 'Ecological modernization and the environmental transition of Europe: between national variations and common denominators', *Journal of Environmental Policy and Planning*, 1: 167–181.

Moore, M. (2003) *Dude, Where's My Country?* London: Penguin.

—— (2001) *Stupid White Men . . . and Other Sorry Excuses for the State of the Nation!* London: Penguin.

Morris, J. (ed.) (2002) *Sustainable Development: Promoting Progress or Perpetuating Poverty?* London: Profile Books.

Munton, R. (1997a) 'Sustainable development: a critical review of rural land-use policy in the UK', in B. Ilbery, Q. Chiotti and T. Rickard (eds) *Agricultural Restructuring and Sustainability*, Oxford: CAB International: 11–24.

—— (1997b) 'Engaging sustainable development: some observations on progress in the UK', *Progress in Human Geography*, 21: 147–63

Murdoch, J. (1997) 'Towards a geography of heterogeneous associations', *Progress in Human Geography*, 21: 321–337.

Murdoch, J. and Marsden, T. (1995) 'The spatialization of politics: local and national actor spaces in environmental conflict', *Transactions of the Institute of British Geographers*, 20: 368–380.

Naess, A. (1994) 'The shallow and deep, long-range ecological movement', in L. Pojman (ed.) *Environmental Ethics: Readings in Theory and Application*, Boston, MA: Jones and Bartlett: 102.

Nawrocki, T. and Szczepanski, M. (1995) 'Environmental bases of ecological consciousness', in J. Wódz (ed.) *The Problems of Ecological Awareness*, Katowice: Wydawnictwo Uniwersytet Śląkiego: 29–37.

North Staffordshire Health Authority (1999) *Working for Health: A Strategy for the Development of Health at Work in North Staffordshire*, Stoke-on-Trent: North Staffordshire Health Authority.

Nowicki, M. (1993) *Environment in Poland: Issues and Solutions*, Dordrecht: Kluwer Academic.

OECD (1994) *Environment for Europe: Environmental Action Programme for Central and Eastern Europe Countries*, Paris: OECD.

Ogborn, M. (2002) 'Writing travels: power, knowledge and ritual on the English East India Company's early voyages', *Transactions of the Institute of British Geographers*, 27: 155–171.

Ohmae, K. (1990) *The Borderless World*, London: Collins.

Oldfield, J.D. (2001) 'Russia, systemic transformation and the concept of sustainable development', *Environmental Politics*, 10: 94–110.

—— (2000) 'Structural economic change and the natural environment in Russia', *Post-Communist Economies*, 12: 77–90.

—— (1999) 'Socio-economic change and the environment – Moscow city case study', *Geographical Journal*, 165: 222–231.

Oldfield, J.D. and Shaw, D.J.B. (2002) 'Revisiting sustainable development: Russian cultural and scientific traditions and the concept of sustainable development', *Area*, 34: 391–400.

O'Riordan, T. (1989) 'Politics, practice and the new environmentalism', in D. Gregory and R. Walford (eds) *Horizons in Human Geography*, London: Macmillan: 395–414.

Osherenko, G. and Young, O. (1989) *The Age of the Arctic: Hot Conflicts and Cold Realities*, Cambridge: Cambridge University Press.

Owens, S. (2003) 'The Royal Commission on Environmental Pollution', in G. Altner, H. Leitschuh-Fecht, G. Michelsen, U.E. Simonis and E.U. von Weizsäcker (ed.) *Jarhbuch Ökologie 2004*, Munich: Verlag C.H. Beck: 96–103.

—— (1994) 'Land, limits and sustainability: a conceptual framework and some dilemmas for the planning system', *Transactions of the Institute of British Geographers*, 19: 439–456.

Owens, S. and Rayner, T. (1999) '"When knowledge matters": the role and influence of the Royal Commission on Environmental Pollution', *Journal of Environmental Policy and Planning*, 1: 7–24.

Paasi, A. (1991) 'Deconstructing regions: notes on the scales of spatial life', *Environment and Planning A*, 23: 239–256.

Pacific Legal Foundation (2005) *About Us*. Online. Available at: http://www.pacificlegal.org/PLFProfile.asp (accessed 13 November 2005).

—— (2004) *Home Page*. Online. Available at: http://www.pacificlegal.org/ (accessed 1 September 2004).

Painter, J. and Philo, C. (1995) 'Spaces of citizenship: an introduction', *Political Geography*, 14: 107–120.

Palmer, J.A. (ed.) (2001) *Fifty Key Thinkers on the Environment,* London: Routledge.

Pavlínek, P. and Pickles, J. (2000) *Environmental Transitions: Transformations and Ecological Defence in Central and Eastern Europe,* London: Routledge.

Pearce, D., Barbier, E. and Makandya, A. (1990) *Sustainable Development: Economics and Environment in the Third World,* Aldershot: Edward Elgar.

Pelling, M. (ed.) (2003) *Natural Disasters and Development in a Globalizing World,* London: Routledge.

—— (2001) 'Natural Disasters?', in N. Castree and B. Braun (eds) *Social Nature,* London: Blackwell: 170–188.

Pepper, D. (1996) *Modern Environmentalism: An Introduction,* London: Routledge.

Peterson, D.J. (1993) *Troubled Lands: The Legacy of Soviet Environmental Destruction,* Boulder, CO: Westview Press.

Pickles, J. (2004) *A History of Spaces: Cartographic Reason, Mapping and the Geo-coded World,* London: Routledge.

Pile, S. (1998) 'What is a city?', in D. Massey, J. Allen and S. Pile (eds) *City Worlds,* London: Routledge.

Pinder, D. (2005) *Visions of the City,* Edinburgh: Edinburgh University Press.

Pope, C. and Rauber, P. (2004) *Strategic Ignorance: Why the Bush Administration is Destroying a Century of Environmental Progress,* San Francisco, CA: Sierra Club Books.

Porritt, J. (1992) *Save the Earth,* London: Dorling Kindersley.

Potter, R., Binns, T., Elliot, J. and Smith, D. (2004) *Geographies of Development,* second edition, Harlow : Prentice Hall.

Pounds, N. (1958) *The Upper Silesian Industrial Region,* Bloomington, IN: Indiana University Publications.

Raban, J. (1974) *Soft City,* London: William Collins.

Raffles, H. (2002) *In Amazonia: A Natural History,* Oxford: Princeton University Press.

Redclift, M. (1987) *Sustainable Development: Exploring the Contradictions,* London: Routledge.

Richards, K. (2004) 'Some ethical grounds for an integrated geography', *Area,* 4: 436–437.

Robbins, P. (2000) 'The practical politics of knowing: state environmental knowledge and local political economy', *Economic Geography,* 76: 126–144.

Roberts, P. (1994) 'Sustainable regional planning', *Regional Studies,* 28: 781–787.

Robinson, N. (ed.) (1993) *Agenda 21: Earth's Action Plan,* New York: Oceana Publications.

Routledge, P. (2003) 'Convergent space: process geographies and grassroots globalization networks', *Transactions of the Institute of British Geographers,* 28: 333–349.

—— (1997) 'The imagineering of resistance: Pollok Free State and the practice of postmodern politics', *Transactions of the Institute of British Geographers,* 22: 359–376.

Rowell, A. (1996) *Green Backlash: Global Subversion of the Environment Movement,* London: Routledge.

Rowland, W.R. (1973) *The Plot to Save the World: The Life and Times of the Stockholm Conference on the Human Environment*, Toronto: Clarke, Irwin & Co.

Sachs, W. (1999) *Planet Dialectics: Explorations in Environment and Development*, London: Zed Books.

Said, E.W. (1995) *Orientalism: Western Conceptions of the Orient*, London: Penguin.

Sale, K. (1980) *Human Scale*, London: Secker & Warburg.

—— (1974) 'Mother of all', in S. Kumar (ed.) *The Schumacher Lectures Vol. 2*, London: Abacus: 244.

Sarul, J. (2000) 'Environmental and social responsibility as the value in modern politics', paper delivered at the meeting of OIKOS International, Warsaw, 10 November. Online. Available at: http://www.oikosinternational.org/publications (accessed 14 March 2005).

Satterthwaite, D. (ed.) (1999) *The Earthscan Reader in Sustainable Cities*, London: Earthscan.

—— (1997) 'Sustainable cities or cities that contribute to sustainable development?', *Urban Studies*, 34: 1667–1691.

Schumacher, E.F. (1973) *Small is Beautiful: A Study of Economics as if People Mattered*, London: Abacus, Sphere Books.

Scott, H.V. (2004) 'A mirage of colonial consensus: resettlement schemes in early Spanish Peru', *Environment and Planning D: Society and Space*, 22: 885–899.

Scott, J.C. (1998) *Seeing Like a State: How Certain Schemes to Improve the Human Condition Have Failed*, New Haven, CT: Yale University Press.

Selva, M. (2004) 'Queen of the Greens', *The Independent*, 9 October: 38–39.

Sennett, R. (1974) *The Fall of Public Man*, Cambridge: Cambridge University Press.

The Sentinel (2000a) 'Heather lives life to the full in Fit City', *Sentinel News Archive*. Online. Available at: http://www.thisisstaffordshire.co.uk (accessed 5 September 2006).

The Sentinel (2000b) 'Shake-up to change sick city to fit city', *Sentinel News Archive*. Online. Available at: http://www.thisisstaffordshire.co.uk (accessed 5 September 2006).

Sharp, J., Routledge, P., Philo, C. and Paddison, R. (2000) *The Entanglements of Power: Geographies of Domination/Resistance*, London: Routledge.

Shiva, V. (1998) *Biopiracy: The Plunder of Nature and Knowledge*, Dartington: Green Books.

—— (1993) *Monocultures of the Mind*, London: Zed Books.

—— (1991a) 'The Green Revolution in the Punjab', *The Ecologist*, 21: 57–60.

—— (1991b) *The Violence of the Green Revolution: Third World Agriculture, Ecology and Politics*, London: Zed Books.

Slocum, R. (2004) 'Polar bears and energy-efficient lightbulbs: strategies to bring climate change home', *Environment and Planning D: Society and Space*, 22: 413–438.

Smith, N. (1993) 'Homeless/global: scaling places', in J. Bird, B. Curtis, T. Putnam, G. Robertson and L. Tickner (eds) *Mapping the Futures: Local Cultures, Global Change*, London: Routledge: 87–119.

—— (1992) 'Geography, difference and the politics of scale', in J. Doherty, E. Graham and M. Malek (eds) *Postmodernism and the Social Sciences*, London: Macmillan: 57–79.

—— (1991) *Uneven Development: Nature, Capital and the Production of Space*, Oxford: Blackwell.

Smith, S. (2002) 'Citizenship', in R. Johnston, D. Gregory, G. Pratt and M. Watts, *The Dictionary of Human Geography*, Oxford: Blackwell: 83–84.

—— (1990) 'Society, space and citizenship: a human geography for the new times', *Transactions of the Institute of British Geographers*, 14: 144–156.

Soja, E. (2004) 'On regions', paper presented to the Institute of Geography and Earth Sciences, University of Wales, Aberystwyth.

—— (1996) *Third Space: Journeys to Los Angeles and Other Real and Imagined Places,* Oxford: Blackwell.

—— (1989) *Postmodern Geographies: The Reinsertion of Space in Critical Social Science*, London: Verso.

Solzhenitsyn, A. (2002) *Gulag Archipelago: 1918–1956* [Parts 1 and 2], London: HarperCollins.

Sparke, M. (2004) 'Political geography: political geographies of globalization (1) dominance', *Progress in Human Geography*, 28(6): 777–794.

Stark, D. (1992) 'The great transformation? Social change in Eastern Europe', *Contemporary Sociology*, 23: 299–304.

Stenning, A. (2005) 'Post-socialism and the changing geographies of everyday life in Poland', *Transactions of the Institute of British Geographers*, 30: 113–127.

Stocking, M. and Perkin, S. (1992) 'Conservation-with-development: an application of the concept in the Usambara Mountains, Tanzania', *Transactions of the Institute of British Geographers*, 17: 337–359.

Storper, M. (1997) *The Regional World: Territorial Development in a Global Economy*, New York: Guilford Press.

Swyngedouw, E. (1997a) 'Neither global or local: "glocalisation" and the politics of scale', in K. Cox (ed.) *Spaces of Globalization: Reasserting the Power of the Local*, New York: Guilford Press: 138–166.

—— (1997b) 'Excluding the Other: the production of scale and scaled politics', in R. Lee and J. Wills (eds) *Geographies of Economies,* London: Arnold 169–176.

Thomas, D.H.L. and Adams, W.M. (1997) 'Space, time and sustainability in the Hadejia-Jama'are wetlands and the Komodugu Yobe basin, Nigeria', *Transactions of the Institute of British Geographers*, 22: 430–449.

Toepfer, K. (2004) *Statement by Klaus Toepfer, Executive Director of the United Nations Environment Programme (UNEP), following the decision by the Russian Duma to Ratify the Kyoto Protocol on climate change* (22 October). Online. Available at: http://www.unep.org/themes/climatechange/infocus/infocus_kyoto.asp (accessed on 18 July 2005).

Toffler, A. (1970) *Future Shock*, London: Pan Books.

Tomeczek, A. (1993) 'Revindicational social movements in the border region: a sociological case study', in M. Szczepanski (ed.) *Dilemmas of Regionalism and the Region of Dilemmas*, Katowice: Uniwersytet Śląski: 162–174.

Turner, B. (2000) 'Review essay: Citizenship and political globalization', *Citizenship Studies*, 4: 81–86.

UNCED (1992) *Agenda 21*. Online. Available at: http://www.unep.org/ Documents.Multilingual/Default.asp?DocumentID=52 (accessed 17 November 2005).

UNCHS/UNEP (2001) *General Information – Sustainable Cities and Local Governance*. Online. Available at: http://www.unchs.org/scp/info/general/ general.htm (accessed 5 November 2005).

United Nations (2003a) *Report of the World Summit on Sustainable Development: Johannesburg, South Africa, 26 August – 2 September 2002*, New York: United Nations.

—— (2003b) *How Will Global Warming Effect my World? A Simplified Guide to the IPCC's Climate Change 2001: Impacts, Adaptation and Vulnerability*, Geneva: United Nations Environment Programme.

—— (1999) *World Urbanization Prospects, The 1999 Revision*, New York: United Nations.

Urban-Klaehn, J. (2005) *A Brief History of the Solidarity Workers' Union*. Online. Available at: http://www.bellaonline.com/articles/art34963. asp (accessed 6 November 2005).

Urry, J. (2000) *Sociology Beyond Societies: Mobilities for the Twenty-first Century*, London: Routledge.

Warrick, R. (1993) 'Slowing global warming and sea level rise: the rough road from Rio', *Transactions of the Institute of British Geographers*, 18: 140–148.

Watts, M. (2001) 'The Progress in Human Geography Lecture, 1968 and all that . . .', *Progress in Human Geography*, 25: 157–188.

—— (1995) 'Colonialism', in R.J. Johnston, D. Gregory and D.M. Smith (eds) *The Dictionary of Human Geography*, Oxford: Blackwell: 75–77.

WCED (World Commission on Environment and Development) (1987) *Our Common Future*, Oxford: Oxford Univesity Press.

Weale, A. (2001) 'The politics of ecological modernization', in J.S. Dryzek and D. Schlosberg (eds) *Debating the Earth: The Environmental Politics Reader*, Oxford: Oxford University Press.

Welsh Assembly Government (2002) *Education for Sustainable Development and Global Citizenship*, Birmingham: ACCAC Publications.

West Midlands Group (on Post-War Reconstruction and Planning) (1948) *Conurbation: A Planning Survey of Birmingham and the Black Country*, London: Architectural Press.

Whitehead, M. (2005) 'Between the marvellous and the mundane: everyday life in the socialist city and the politics of the environment', *Environment and Planning D: Society and Space*, 23: 273–294.

—— (2003a) '(Re)Analysing the sustainable city: nature, urbanization and the regulation of socio-environmental relations in the UK', *Urban Studies*, 40: 1183–1206.

—— (2003b) 'Love thy neighbourhood: rethinking the politics of scale and Walsall's struggle for neighbourhood democracy', *Environment and Planning A*, 35: 277–300.

—— (2003c) 'From moral space to the morality of scale. The case of the sustainable region', *Ethics, Place and Environment*, 6: 235–257.

—— (2003d) 'Interactive learning and the politics of sustainable development', *Environmental Education*, Autumn: 31–35.

—— (2000) 'Regional devolution and the rise of the sustainable region: lessons from the English regional sustainable development fora', *Regions: The Newsletter of the Regional Studies Association*, 299: 11–14.

Whitehead, M., Jones, M. and Jones, R. (2006) 'Spatializing the Ecological Leviathan: state strategies and the production of regional natures in the UK', *Geografiska Annaler: Series B, Human Geography*, 88b: 49–66.

World Bank (2000) *World Development Report 2000–2001: Attacking Poverty*, Oxford: Oxford University Press.

—— (1986) *World Development Report*, Washington, DC: World Bank.

World Resources Institute (2002) *Background Paper: Center for Sustainable Transport in Mexico City*, Washington, DC: World Resources Institute.

WWF (2004) *Balancing Development with Biodiversity Conservation in the Barents Sea: WWF's Barents Sea Ecoregion Programme*. Online. Available at: http://www.panda.org/about_wwf/where_we_work/arctic/what_we_do/marine/barents/publications/index.cfm (accessed 5 September 2006).

—— (2001) *The Barents Sea Ecoregion – A Biodiversity Assessment*. Online. Available at: http://www.panda.org/about_wwf/where_we_work/arctic/what_we_do/marine/barents/publications/index.cfm (accessed 5 September 2006).

Yapa, L. (1996) 'Improved seeds and constructive scarcity', in R. Peet and M. Watts (eds) *Liberation Ecologies: Environment, Development and Social Movements*, London: Routledge: 69–85.

Yearley, S. (1996) *Sociology, Globalization, Environmentalism*, London: Sage.

INDEX